A Field Guide to

Giant Clams

of the Indo-Pacific

A Field Guide to

Giant Clams

of the Indo-Pacific

Written by

Mei Lin Neo
National University of Singapore, Singapore

Illustrated and Designed by

Adel Lee

World Scientific

NEW JERSEY · LONDON · SINGAPORE · BEIJING · SHANGHAI · HONG KONG · TAIPEI · CHENNAI · TOKYO

Published by

World Scientific Publishing Co. Pte. Ltd.
5 Toh Tuck Link, Singapore 596224
USA office: 27 Warren Street, Suite 401-402, Hackensack, NJ 07601
UK office: 57 Shelton Street, Covent Garden, London WC2H 9HE

National Library Board, Singapore Cataloguing in Publication Data
Name(s): Neo, Mei Lin. | Lee, Adel, illustrator, book designer.
Title: A field guide to giant clams of the Indo-Pacific /
    written by Mei Lin Neo ; illustrated and designed by Adel Lee.
Description: Singapore : World Scientific Publishing Co. Pte. Ltd., [2023]
Identifier(s): ISBN 978-981-12-7417-6 (hardcover) |
    ISBN 978-981-12-7418-3 (ebook for institutions) |
    ISBN 978-981-12-7419-0 (ebook for individuals)
Subject(s): LCSH: Giant clams--Indo-Pacific Region. |
    Giant clams--Indo-Pacific Region--Identification.
Classification: DDC 594.41774--dc23

British Library Cataloguing-in-Publication Data
A catalogue record for this book is available from the British Library.

Disclaimer: The views expressed are those of the author (ML Neo) and do not necessarily reflect the views of The Pew Charitable Trusts.

For any available supplementary material, please visit
https://www.worldscientific.com/worldscibooks/10.1142/13349#t=suppl

Table of Contents

Meet "Mama Jong", my first wild giant clam sighting in Singapore waters. This is a Fluted Giant Clam (Tridacna squamosa), one of the two native species extant in southern Singapore. Locally, giant clam population numbers have declined significantly due to fishing, habitat loss, sedimentation, and poor water quality. Still, a small population persists, and there are efforts to rehabilitate populations with cultured individuals.

Preface

I can still fondly remember my first encounter with a wild giant clam in Singapore waters. "Giant clam!" A volunteer had hollered for me across the vast intertidal reef flat of Pulau Jong, one of the last untouched islands in southern Singapore, as I shuffled through the shallow tidal pools, eager to see the animal. The first thought that came to my mind was that it was indeed a 'giant' clam, considering that the only other bivalves that I had seen before were mussels and oysters. At that time, I was bombarded with questions from the volunteers asking me if I could share more about the giant clam, as they knew I was studying them for my undergraduate project. I did my best to share what little I knew based on the papers I read as they listened intently to me. That was my first attempt at science communication – conveying scientific information into bite-sized facts to a lay audience.

The giant clam is a fascinating species once you get to know it better. Even though the animal may appear inanimate at first glance, my research and observations have shown otherwise! Some exciting behaviours we have studied include the swimming ability of microscopic larvae, the 'walking' behaviour of juvenile clams, and the spawning of adult clams. These behaviours facilitate my understanding of how a giant clam and its babies could survive in the open sea. I will admit that I am incredibly biased and empathetic towards them because of their conservation plight. More importantly, the giant clam had shown me what it means not to give up easily. As with many marine species, the giant clam also starts its life as a tiny microscopic larva in the open sea. It is an immense challenge for a small larva to navigate the oceans, to avoid being eaten, to search for suitable shelters, and to find food. With today's additional human pressures, their challenge to survive is insurmountable. Despite the increasing odds against them, the giant clam persists in many parts of the Indo-Pacific region, albeit in lower numbers compared to the distant past.

In my career, I have found marine conservation work to be fulfilling yet also very challenging. As a marine ecologist, I primarily focused my research on the interactions of marine organisms with their environment to discover new information that can inform decisions on managing natural resources. Science communication has also been critical in my efforts to share scientific information widely and raise awareness of the endangered giant clams for marine conservation initiatives. Even as I had stepped out of my academic comfort zone into the realm of science communication, my effort had been missing an important puzzle piece: how to better relate science to the public. I began to rethink what I knew of marine conservation and re-learn that it is not only just about how we manage biodiversity, but it is equally, if not more, important to recognise that people's local knowledge, values, and behaviours can influence decisions related to wildlife conservation. I revamped my approach from conveying my knowledge in a unidirectional way to having a bidirectional conversation with my audience. More importantly, I listened to them to find out what knowledge is needed to strengthen the impetus for the marine conservation of giant clams.

The result? I wrote this entire field guide to giant clams from scratch! With a growing interest from the community asking about giant clam species identification and their basic biology, the motivation for writing this field guide is to provide a valuable and easy-to-read resource for academic researchers, biology students, the SCUBA diving community, managers of marine resources, as well as the public. This field guide contains the most up-to-date information about the giant clams, supplemented with high-quality illustrations and photographs. It provides a taxonomic key for identifying giant clams with detailed information on each species. This book also tries to answer the many conservation questions and issues raised by the community, biologists and conservation managers. The interpretations of information here are my own, and it is alright to disagree with me. Nevertheless, I sincerely hope it is helpful to you all!

Finally, I am glad that I did not give up on my mission to champion the conservation of giant clams, even when the going got tough. After spending more than a decade researching giant clams, I want to do right by them by amplifying my knowledge about them in this field guide. I hope readers will enjoy learning the charming sides of these fantastic marine bivalves and draw them into the world of these gentle underwater giants...

Mei Lin Neo (also known as the "Giant Clam Girl")
Tropical Marine Science Institute
National University of Singapore

©Victor Tang

1. Introduction

Giant clams are enormous bivalves often seen anchored in the shallow coastal habitats such as seagrass meadows and coral reefs across the Indo-Pacific region. Also referred to as 'tridacnids' or 'tridacnines', these sessile molluscs are highly conspicuous because of their exposed and vibrantly coloured outer mantle tissue in their shells. Giant clams boast an average lifespan of half a century or older! These large clams play critical ecological roles in coral reef ecosystems and provide a source of nutrition and income for coastal communities. All species of giant clams are considerably larger than most other bivalves. The largest of giant clams is the *Tridacna gigas* (also known as the true giant clam), which is capable of growing up to 1.2 metres (4 feet) across and weighing over 250 kilograms (500 pounds) – making it the largest mollusc on Earth. Even though the smallest species of giant clams, *Tridacna crocea* (also known as the burrowing giant clam), reaches up to 15 cm long, it is still larger than most bivalves.

Globally, giant clams are generally found in the shallow coastal waters from South Africa to the Pitcairn Islands (32°E to 128°W) and from southern Japan to Western Australia (24°N to 15°S). However, the extent of geographic distributions differs among the 12 giant clam species. The two species with the most widespread geographic distributions are *Tridacna maxima* and *Tridacna squamosa*. At the same time, the *Tridacna gigas*, *Tridacna derasa*, *Tridacna noae*, *Tridacna crocea* and *Hippopus hippopus* have intermediate geographic ranges. The remaining species including *Tridacna*

mbalavuana, Tridacna squamosina, Tridacna rosewateri, Tridacna elongatissima and *Hippopus porcellanus,* have only been reported from a few locations across the Indo-Pacific region.

The species richness (i.e., the number of species in a specific area) of giant clams also varies considerably throughout the Indo-Pacific region. The Central Indo-Pacific is a hotspot for giant clam diversity since eight species are reportedly present in the marine provinces of the Tropical Northwestern Pacific, Western Coral Triangle and Tropical Southwestern Pacific. As we extend outwards from the Central Indo-Pacific, the number of giant clam species encountered is fewer in the remaining marine provinces. The observed patterns of differing species richness across the region may be attributed to several factors, such as the availability and distribution of suitable habitats and environmental conditions to host certain species of giant clams and the flexibility a species may have to prolong settlement and metamorphosis by 'walking' to increase the geographic range of the species.

Broadly, the giant clams make a wide range of ecological contributions to coral reefs and are vital human coastal resources. However, the heavy exploitation of these large bivalves has elevated their threatened status through the Indo-Pacific region. The impending loss of giant clams has consequently provided much-needed attention to the Convention on International Trade in Endangered Species of Wild Fauna and Flora (CITES) and the International Union for Conservation of Nature (IUCN).

Geographic Distributions of Giant Clam Species Across the Indo-Pacific Region

EG	Egypt	**HK**	Hong Kong
JO	Jordan	**TW**	Taiwan
YE	Yemen	**JP**	Japan
KE	Kenya	**PH**	Philippines
TZ	Tanzania	**PW**	Palau
MZ	Mozambique	**TP**	East Timor
SA	South Africa	**PG**	Papua New Guinea
MG	Madagascar	**MP**	Northern Mariana Islands
MU	Mauritius	**FM**	States of Micronesia
CCA	Cargados Carajos Archipelago	**MH**	Marshall Islands
SMB	Saya de Malha Bank	**SB**	Solomon Islands
SC	Seychelles	**KI**	Republic of Kiribati
IO	British Indian Ocean Territory	**PF**	French Polynesia
IN	India	**PN**	Pitcairn Islands
LK	Sri Lanka	**CK**	Cook Islands
CX	Christmas Island	**NU**	Niue
MY	Malaysia	**TO**	Tonga
MM	Myanmar	**FJ**	Fiji
VN	Viet Nam	**NC**	New Caledonia
ID	Indonesia	**QLD**	Queensland, Australia
CN	People's Republic of China	**WA**	Western Australia, Australia

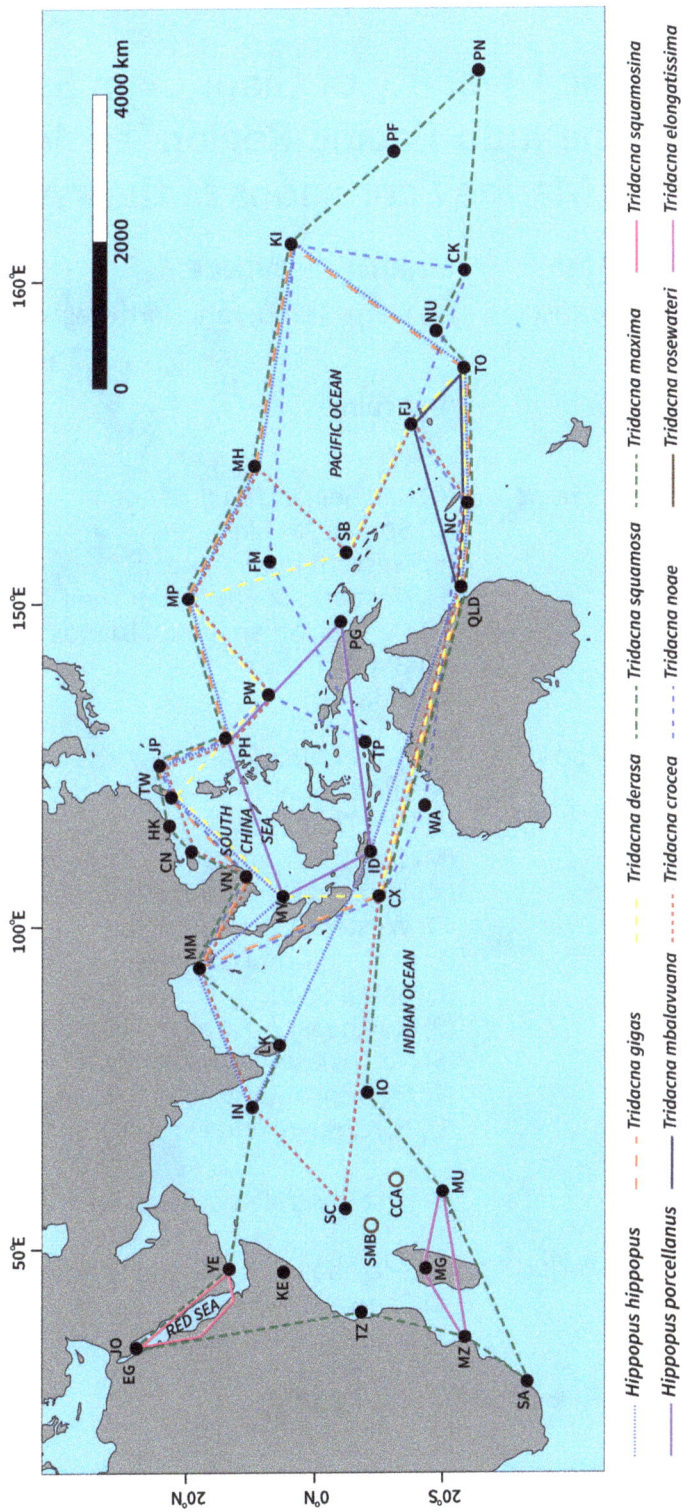

Reproduced from Neo et al. (2017) with updates.

Legend:

- ······· *Hippopus hippopus*
- ——— *Hippopus porcellanus*
- – – – *Tridacna gigas*
- ——— *Tridacna mbalavuana*
- – – – *Tridacna derasa*
- – – – *Tridacna crocea*
- – · – · *Tridacna squamosa*
- ——— *Tridacna noae*
- ——— *Tridacna squamosina*
- – · – · *Tridacna maxima*
- ——— *Tridacna rosewateri*
- ——— *Tridacna elongatissima*

9

Taxonomic Diversity of Giant Clam Species Across the Indo-Pacific Region, as defined by the Marine Ecoregions of the World

MARINE REALM	MARINE PROVINCE
Temperate Northern Pacific	(A) Warm Temperate Northwest Pacific
Temperate Southern Africa	(B) Agulhas
Western Indo-Pacific	(C) Red Sea and Gulf of Aden (D) Somali/Arabian (E) Western Indian Ocean (F) West and South Indian Shelf (G) Central Indian Ocean Islands (H) Bay of Bengal (I) Andaman
Central Indo-Pacific	(J) South China Sea (K) Sunda Shelf (L) Java Transitional (M) South Kuroshio (N) Tropical Northwestern Pacific (O) Western Coral Triangle (P) Eastern Coral Triangle (Q) Sahul Shelf (R) Northeast Australian Shelf (S) Northwest Australian Shelf (T) Tropical Southwestern Pacific (U) Lord Howe and Norfolk Islands
Temperate Australasia	(V) East Central Australian Shelf
Eastern Indo-Pacific	(W) Marshall, Gilbert, and Ellis Islands (X) Central Polynesia (Y) Southeast Polynesia

Source of information: Spalding et al. (2007).

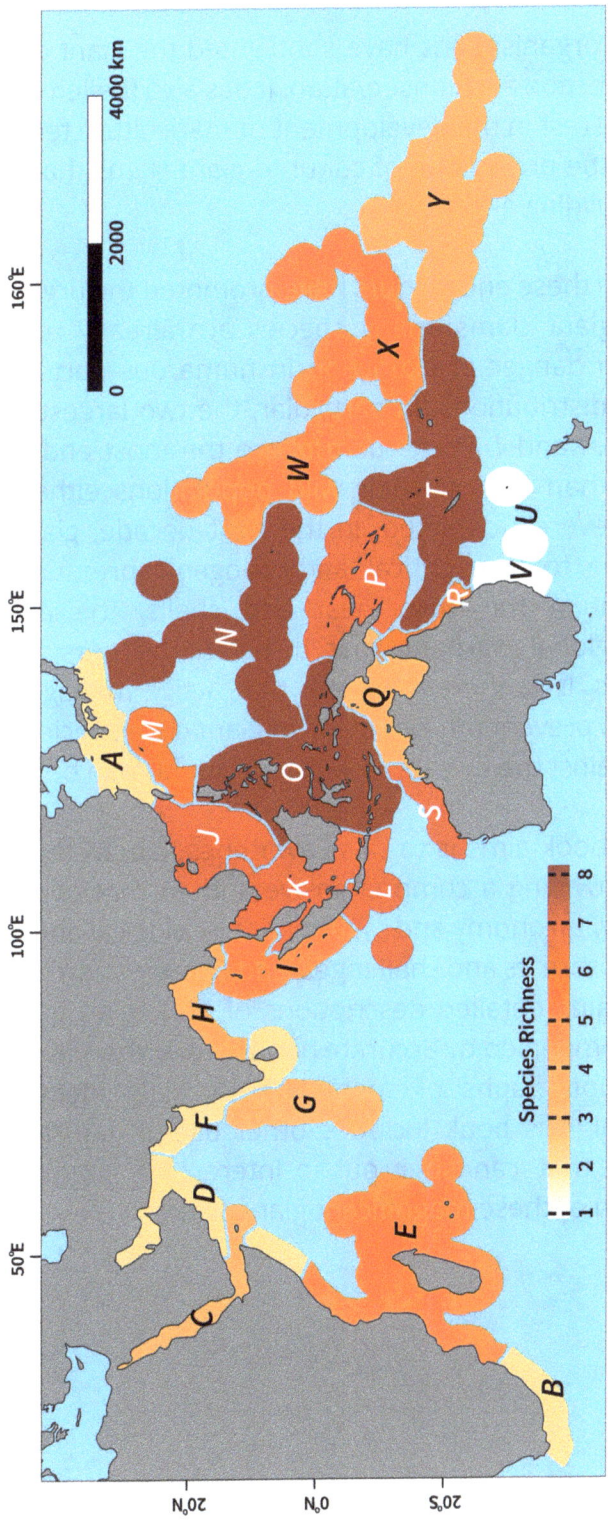

Reproduced from Tan et al. (2022).

These large organisations have spotlighted the giant clams for greater protection and conservation. It has also fuelled a greater scientific interest in the development of mariculture techniques to improve the production of cultured giant clams that may be used for rewilding efforts.

While these endeavours have promoted the urgent need to protect giant clams, many species are already reportedly extinct or in danger of extinction in numerous parts of their geographic distributions. In particular, the two largest species, *Tridacna gigas* and *Tridacna derasa*, are the most endangered, with more than 50% of their wild populations either locally extinct or severely depleted. In the next decade, giant clams will continue to face critical anthropogenic pressures from overexploitation for their meat and shells, the aquarium trade and global warming that threaten the persistence of their species. It is, therefore, clear that active management is necessary to prevent the extinction of giant clam species as they come up against the threats associated with human behaviours.

This book aims is to raise awareness about these giant clams by providing a comprehensive and current overview of their biology, taxonomy and systematics, ecological and cultural significance, threats and challenges, and conservation solutions. It also contains detailed descriptions of 12 known giant clam species accompanied by accurate hand-drawn shell illustrations and live photographs of specimens for field identification. In additional, this book includes other useful natural history information that can spur public interest to champion the conservation of these magnificent giant clams.

With its brightly coloured and speckled fleshy lips peeking out of its massive shells, you might think a giant clam would be hard to miss. After all, it is, in reality, nature's couch potato, being entirely sessile in adulthood, meaning that it simply cannot move (much). Yet these large bivalves tuck themselves so artfully into their habitat that predators, not to mention divers and snorkellers, often drift past without noticing them.

©Victor Tang

2. Taxonomy and Systematics

Giant clams are morphologically derived true cockles (also referred to as cardiids) from the family of Cardiidae. These bivalves have evolved an obligate symbiotic relationship with the photosynthetic dinoflagellates. Previously, the giant clams were considered a distinct family under Tridacnidae Lamarck, 1819 within the former order Venerida Gray, 1854. Still, subsequent morphological and phylogenetic evidence gathered over the past two decades has confirmed that they are a monophyletic subgroup in the family Cardiidae Lamarck, 1809. Even though some scientists have argued to maintain 'Tridacnidae' at the family level rank based solely on its highly unique morphology, the most widely accepted and recognised taxonomic ranks for giant clams, according to the World Register of Marine Species (WoRMS), are summarised in the following table.

In 1965, renowned malacologist Joseph Rosewater (1928-1985) wrote a comprehensive review on the taxonomic status of six giant clam species in *"The family Tridacnidae in the Indo-Pacific"*. Before the publication of Rosewater's seminal paper, several others before him had attempted to review and revise the descriptions of giant clam species, but many of their described species names have been synonymised. On the other hand, the six species recognised by Rosewater have remained valid to this day. After nearly five decades since Rosewater's work, the number of giant clam species has exactly doubled with descriptions of new species and the resurrection of previously synonymised names. The species taxonomy and systematics for the giant clams have also shifted from heavy reliance on

Taxonomic Classifications for the Giant Clams

Phylum:	Mollusca
Class:	Bivalvia Linnaeus, 1758
Order:	Cardiida A. Férussac, 1822
Family:	Cardiidae Lamarck, 1809
Subfamily:	Tridacninae Lamarck, 1819
Genera:	*Hippopus* Lamarck, 1799 *Tridacna* Bruguière, 1797
Subgenera:	*Tridacna (Tridacna)* Bruguière, 1797 *Tridacna (Persikima)* Iredale, 1937 *Tridacna (Chametrachea)* Herrmannsen, 1846
Species:	*Hippopus hippopus* (Linnaeus, 1758) *Hippopus porcellanus* Rosewater, 1982 *Tridacna (Tridacna) gigas* (Linnaeus, 1758) *Tridacna (Tridacna) mbalavuana* Ladd, 1934 *Tridacna (Persikima) derasa* (Röding, 1798) *Tridacna (Chametrachea) crocea* Lamarck, 1819 *Tridacna (Chametrachea) squamosa* Lamarck, 1819 *Tridacna (Chametrachea) noae* (Röding, 1798) *Tridacna (Chametrachea) maxima* (Röding, 1798) *Tridacna (Chametrachea) rosewateri* Sirenko & Scarlato, 1991 *Tridacna (Chametrachea) squamosina* Sturany, 1899 *Tridacna (Chametrachea) elongatissima* Bianconi, 1856

Source of information: World Register of Marine Species (WoRMS).

morphological characters alone to differentiate species before the 1990s to using molecular tools combined with morphological traits in delimiting species in the late 2000s. The latter has been especially useful in identifying highly cryptic species. Given the high variation and plasticity typically observed in their shell and mantle morphologies, delimiting contemporary giant clam species should include genetic data to support comparisons.

Timeline of Species Discoveries

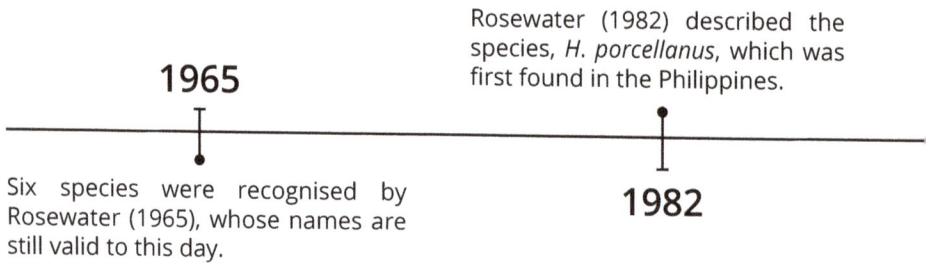

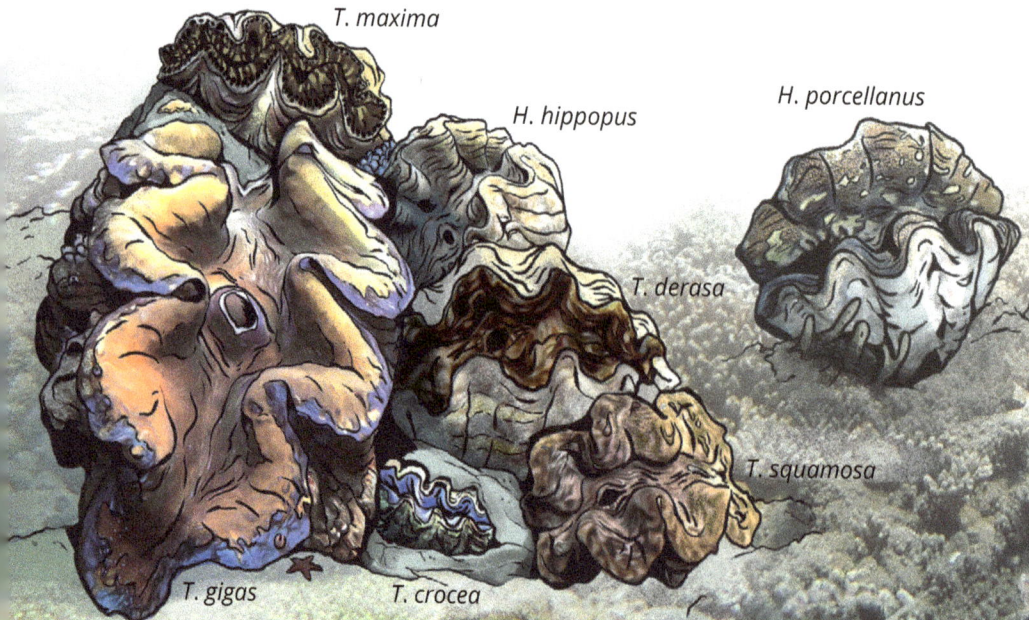

1965

Six species were recognised by Rosewater (1965), whose names are still valid to this day.

Rosewater (1982) described the species, *H. porcellanus*, which was first found in the Philippines.

1982

T. maxima

H. hippopus

H. porcellanus

T. derasa

T. squamosa

T. gigas

T. crocea

Considering the increase in species diversity and the ease of accessibility to genetic material of rare or uncommon species, a reappraisal of the phylogenetic relationships between giant clam species is urgently needed. Edwin Tan and his colleagues undertook and published this work, where they had reconstructed the most comprehensive and data-supported phylogeny of this subfamily using genome

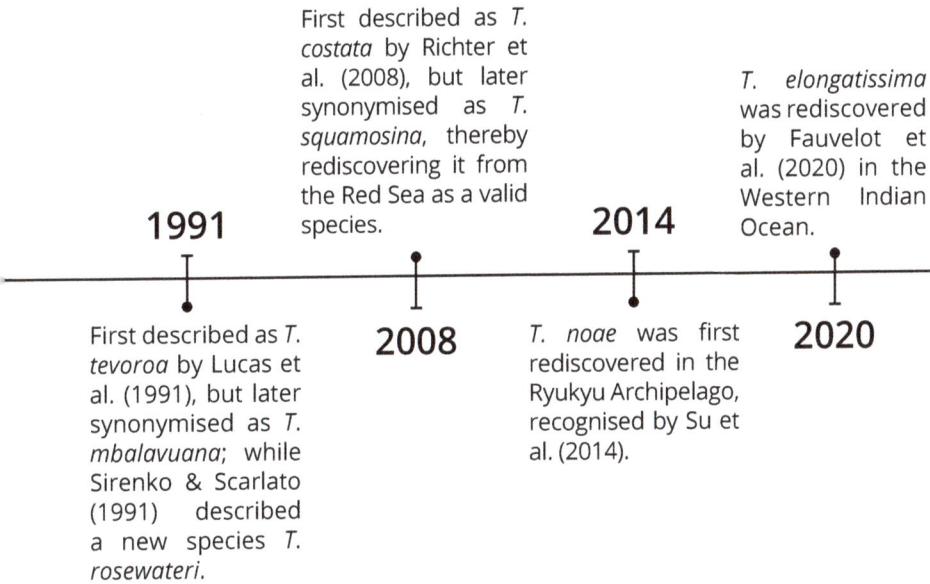

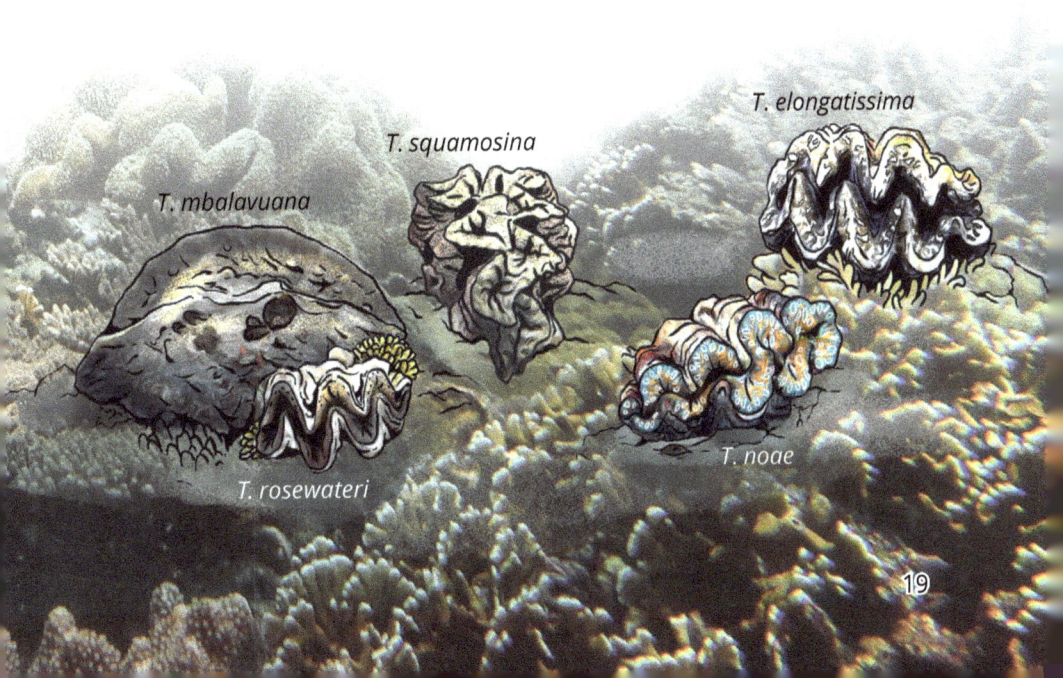

1991

First described as *T. tevoroa* by Lucas et al. (1991), but later synonymised as *T. mbalavuana*; while Sirenko & Scarlato (1991) described a new species *T. rosewateri*.

First described as *T. costata* by Richter et al. (2008), but later synonymised as *T. squamosina*, thereby rediscovering it from the Red Sea as a valid species.

2008

2014

T. noae was first rediscovered in the Ryukyu Archipelago, recognised by Su et al. (2014).

T. elongatissima was rediscovered by Fauvelot et al. (2020) in the Western Indian Ocean.

2020

T. mbalavuana

T. squamosina

T. elongatissima

T. rosewateri

T. noae

skimming. An emerging technique, this genome skimming method involves the shallow sequencing of a target genome to produce short fragments of DNA, known as genome skims. These genome skims typically contain information used for phylogenomic analyses across a wide range of evolutionary divergences. Compared to other sequencing approaches, genome skimming provides an affordable and robust strategy in assembling target genomes for phylogenetic studies.

In this latest study, the phylogenetic assessment had incorporated 12 extant species of giant clams and used 16 gene markers for analyses, which included 15 mitochondrial

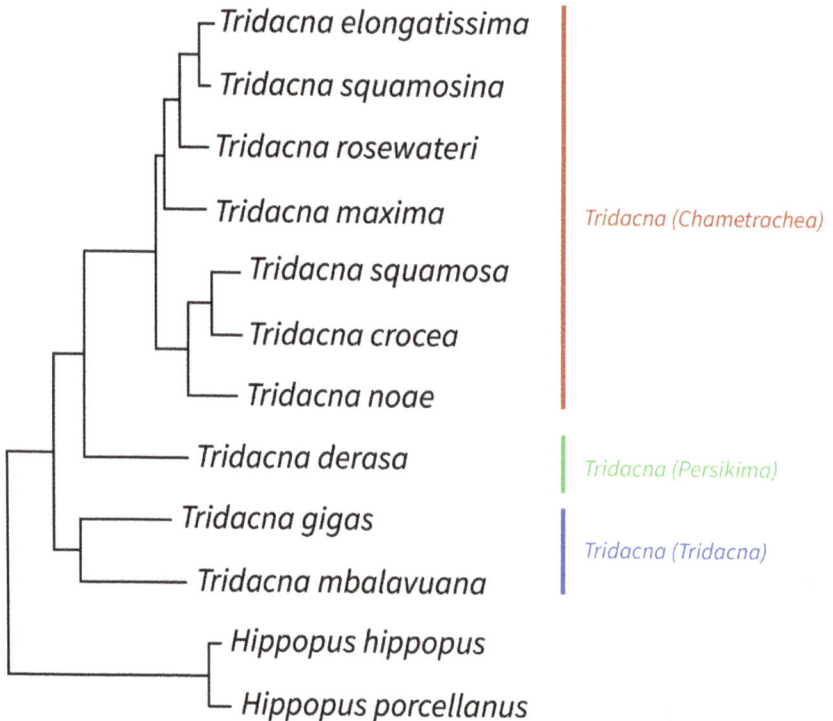

The consensus phylogenetic tree for the giant clam species in the subfamily Tridacninae obtained from Maximum Likelihood and Bayesian analyses. The respective subgenera are presented here in coloured text. Reproduced from Tan et al. (2022).

protein-coding genes and one nuclear 18S rRNA gene. This endeavour is different from previous phylogenetic studies that typically used one to few genetic markers, which only sometimes provide sufficient resolution to confirm species relationships. Furthermore, representations from rarer species such as *Hippopus porcellanus*, *Tridacna mbalavuana*, *Tridacna rosewateri* and *Tridacna squamosina* could additionally sort out the species relationships.

The resulting phylogeny based on a concatenated 16-gene dataset reveals highly supported relationships within and between the three subgenera, *Tridacna (Tridacna)*, *Tridacna (Persikima)* and *Tridacna (Chametrachea)*. The study found that including a larger gene dataset dramatically improves the support and validity of the subfamily's phylogeny. It also yielded novel findings such as the well-supported relationship between *Tridacna gigas* and *Tridacna mbalavuana* within subgenus *Tridacna (Tridacna)* despite their vastly different morphologies, as well as the current species relationships among *Tridacna elongatissima*, *Tridacna squamosina*, *Tridacna rosewateri* with *Tridacna maxima* as the former three species were included in this assessment for the first time. The availability of this comprehensive phylogenetic dataset will serve as the initial baseline of Tridacninae relationships to support future studies examining their systematics, ecology and conservation.

"Excuse Me - Are you a Giant Clam?"

If you think differentiating giant clam species is challenging, wait until you meet these 'imposters'! Check out how these other marine species may become mistaken for giant clams.

1. Cockles (Family Cardiidae)

A large and diverse group of bivalves, which includes the giant clams. Found worldwide in shallow tropical to temperate waters. Cardiidae shells are typically rounded to trigonal or quadrangular in outline and most often dorsoventrally elongated. Giant clams are an exception where their large shells are suboval to fan-shaped in outline and are anteroposteriorly elongated. The *Corculum cardissa s.s.*, some *Fragum* species, and all the Tridacnines have photosynthetic endosymbionts.

2. Honeycomb Oysters (Family Gryphaeidae)

Medium to large bivalves with a size range up to 30 cm (11.8 inches). Found in tropical and subtropical waters. Gryphaeidae shells are usually inequivalve with a circular or oval shape. These marine or estuarine animals cement themselves to hard surfaces. May be mistaken for giant clams due to their interlocking shell margins, but the shell margins of giant clams are usually rounder with gaping valves.

3. Thorny Oysters (Family Spondylidae)

Medium bivalves with a size range up to 20 cm (7.8 inches). Found worldwide in tropical and subtropical waters from shallow to very deep. Spondylidae shells are inequivalve with a general subcircular shape. These bivalves live permanently attached to hard surfaces. May be mistaken for juvenile *Hippopus* spp. due to prominent spines on their valves, but the spines in *Hippopus* spp. are much shorter.

4. Pen Shell (Family Pinnidae)

Large marine bivalves with a size range up to 90 cm (35.4 inches). Found worldwide in shallow tropical and subtropical waters. Pinnidae shells are long and trigonal in outline. The anterior ends of their shells are embedded in sediment, which exposes the broader, gaping posterior opening above the sediment-water interface. May be mistaken for giant clams due to their large size.

5. Coral Scallop (*Pedum spondyloideum*)

It is a boring species of pectinid bivalve associated closely with scleractinian and hydrozoan corals. Found worldwide in shallow tropical and subtropical waters. It can reach up to 8 cm (3.1 inches) in size, with rounded, oval shells with smooth shell margins. These animals live fully embedded in shallow depressions. May be mistaken for juvenile *Tridacna crocea* due to its boring behaviour in coral rocks.

6. Flatworms (*Pseudobiceros* spp.)

Marine flatworms are unsegmented and soft-bodied invertebrates. Found worldwide but most common in tropical oceans. These brightly coloured animals can reach up to 10 cm (4 inches) in size. Flatworms have a flat and smooth dorsal surface with ruffled edges used for locomotion and swimming. May be mistaken for juvenile *Tridacna crocea* due to their brightly coloured ruffled edges.

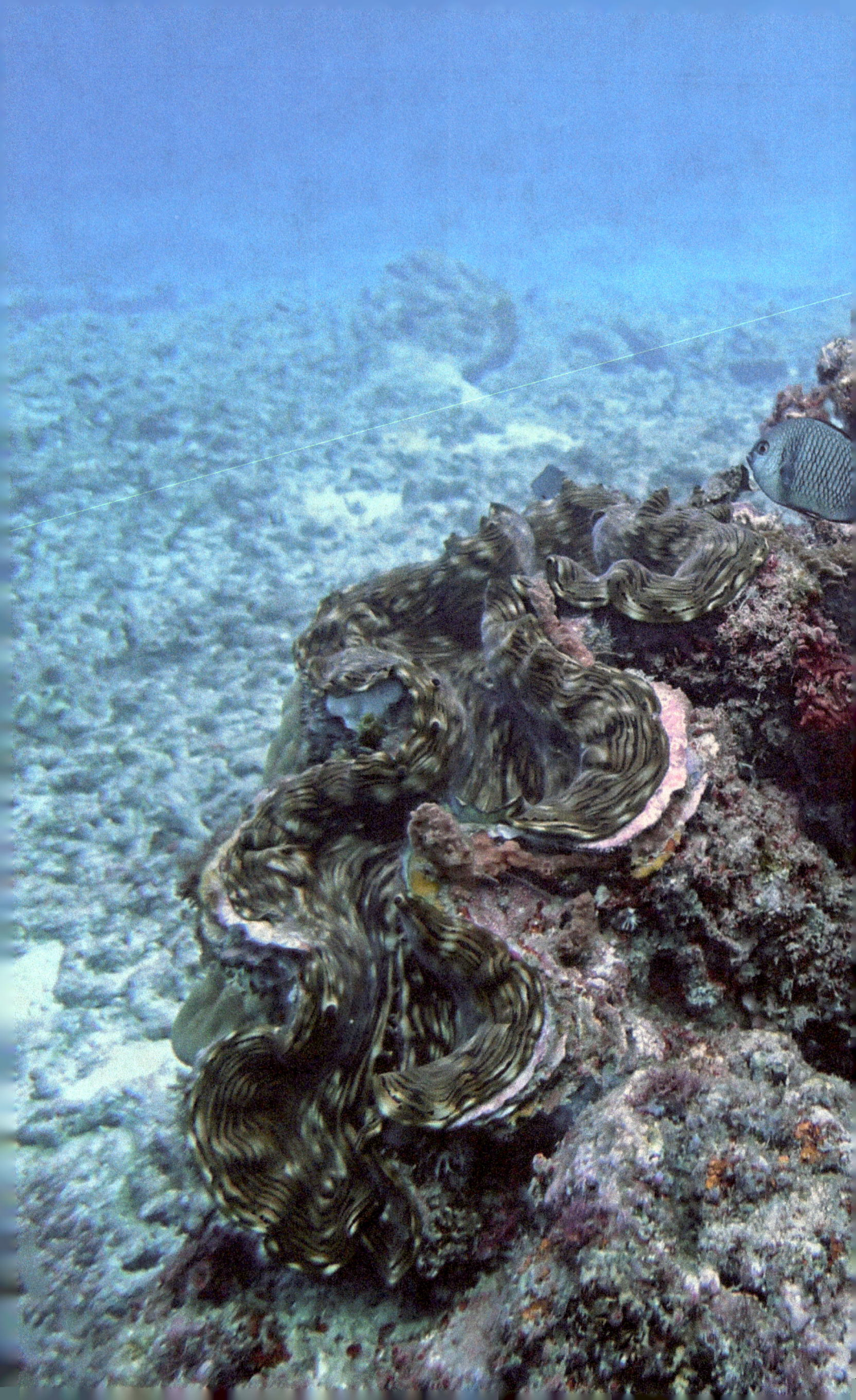

3. Biology

Anatomy

The giant clam is a bivalve mollusc, which means the animal is fully enclosed by two shell valves attached via a hinge ligament made of an elastic protein. The shell is also held together by a large, central adductor muscle between the two valves that keep them closed. In contrast to other bivalves, giant clams orient themselves with their hinge surface facing ventrally towards the seabed, while their outer mantle is facing dorsally towards the light. Giant clam shells are entirely aragonite, a mineral form of calcium carbonate, making them useful as biorecorders for past climate and environmental reconstructions. Shell valves are typically suboval or fan-shaped in outline. The external surfaces of valves generally are greyish-white, with a primary radial sculpture of strongly convex and rib-like folds. In contrast, the secondary radial ornamentation consists of evenly spaced riblets interrupted by fine concentric growth lines. In the species of subgenus *Tridacna* (*Chametrachea*), the ribs of shell valves also possess scute-like projections.

The mantle (sometimes referred to as siphonal mantle) encloses the soft organs, including the heart, kidney, gonads and ctenidia (gills), and has the important function of secreting the shell. It has three openings: the incurrent and excurrent siphonal openings and the byssal orifice. The incurrent siphon is the larger opening that may be lined with guard tentacles in some species, while the excurrent siphon is smaller and cone-shaped. Both anatomical structures facilitate water movement

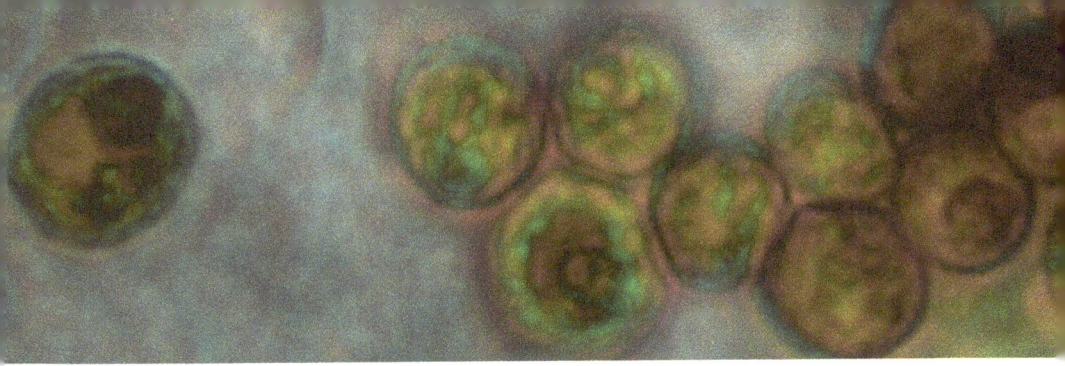

in and out of the giant clam for feeding, breathing, and expelling of waste. The byssal orifice is next to the umbo on the ventral side (underside of shell). In young clams, the foot extends out from this opening and makes contact with the surface to 'walk' in search of suitable habitats. The byssus, produced by the gland in the foot, then anchors the clams to the seabed. Some species, such as *Tridacna crocea*, *Tridacna maxima* and *Tridacna squamosa*, retain byssal attachment throughout life, whereas the foot of *Tridacna gigas* and *Tridacna derasa* atrophy after reaching a specific body size.

The enlarged siphonal mantle covers the area between the valves and usually extends beyond the upper shell margins when the clam is open. Notably, only the mantles of *Hippopus* spp. and *Tridacna mbalavuana* do not extend beyond the edge of their shells. This thick muscular tissue is packed densely with the photosynthetic dinoflagellate algae from the family Symbiodiniaceae. The brown colours are usually attributed to the high densities of algae. The mantle's surface is also covered by a layer of specialised cells called iridocytes, which reflect light and give the clams their vibrant and colourful mantles in blues, greens or golds. In addition, all giant clam species, except *Hippopus* spp. and *Tridacna mbalavuana*, possess hundreds of small pinhole eyes, known as hyaline organs. These 'eyes' are light-sensitive, which allow the giant clams to detect changes in light levels. Hyaline organs are usually found adjacent to the mantle's margins, but may also be scattered over non-marginal areas of the siphonal mantle.

Hyaline organs

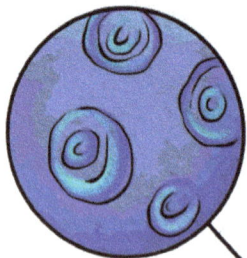

Guard tentacles

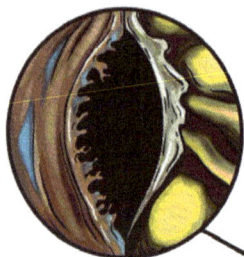

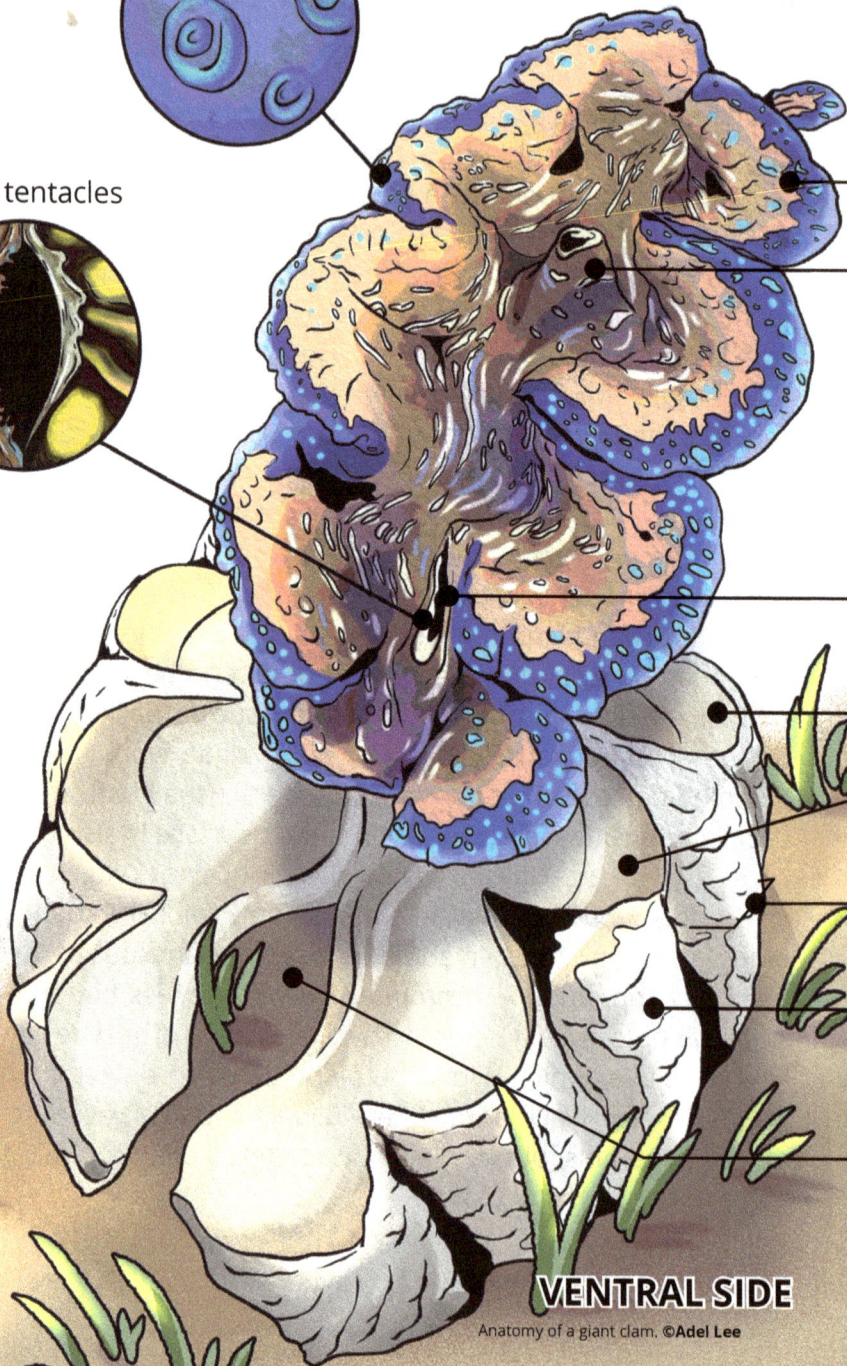

DORSAL SIDE

VENTRAL SIDE

Anatomy of a giant clam. ©**Adel Lee**

28

Anatomy of a Giant Clam

Shell symmetry

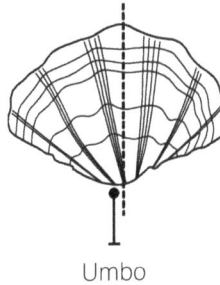

Umbo

Shell symmetry is equilateral if umbo is in the middle, and inequilateral if umbo leans on one side.

Siphonal mantle

Excurrent siphon

Shell sculpture

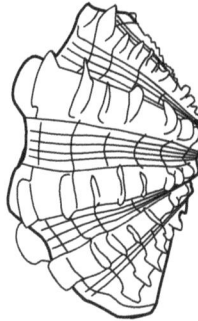

Shell sculpture is the external texture and patterns of its shell valve, and overlaps with growth bands.

Incurrent siphon

Valve margins

Primary radial sculpture

Secondary radial ornamentation

Byssal orifice

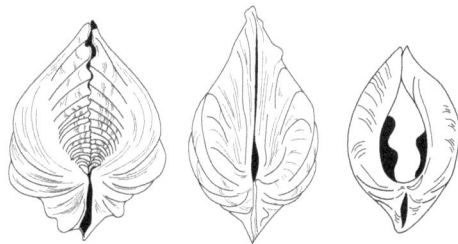

Interlocking   Narrow   Wide

Byssal orifice opening

Byssal orifice is the opening on the underside of the shell. Its size varies across the giant clam species, often used as a distinguishing trait.

Feeding

Giant clams obtain nutrients through two pathways: filter-feeding and photosynthesis. The larvae, juveniles and adults are efficient filter-feeders that can capture suspended particulate food such as phytoplankton, micronutrients and trace metals via their ctenidia. These bivalves also get their 'solar-power' capability from symbiosis with the photosynthetic dinoflagellates (also referred to as endosymbionts). The latter provides the animal hosts with an additional source of nutrition other than the ingestion of food items in exchange for a sheltered environment. Only a few extant bivalve molluscs have maintained symbioses with Symbiodiniaceae, including giant clams, the heart cockles (genus *Corculum*), and true cockles (genus *Fragum*) that are all members of the family Cardiidae. In the case of giant clams, they have developed several morphological adaptations to maximise their photosymbiotic lifestyle. Firstly, the epibenthic giant clams reside above the seabed that allows them to be fully exposed to light. In addition, the anatomical body plan of giant clams is highly modified, where the siphonal mantle is greatly enlarged and exposed upwards towards the light between their widely gaping valves. The enlarged siphonal mantle contains high densities of Symbiodiniaceae algae that reside intercellularly in a zooxanthellae tubular system. This complex network of tubules carrying endosymbionts is also located underneath a layer of light-scattering iridocyte cells on the surface of the siphonal mantle. All these features allow for the maximum light capture by these endosymbionts through increased surface area exposed to light.

Studies have confirmed that the host giant clams depend heavily on their endosymbionts to obtain the bulk of their carbon and nitrogen requirements for growth and metabolism, even though they have functional ctenidia and digestive systems typical of heterotrophic bivalves. Genetic

Before feeding (t = 0h)

After feeding (t = 3h)

Giant clams filter feed by drawing in seawater through their incurrent siphon and filter out phytoplankton and sediments, thus cleaning the water. Here we show the before and after feeding of a single *Tridacna crocea* individual with 300,000 cells of *Tisochrysis* microalgae, demonstrating its efficient ability to clear the water! ©Mei Lin Neo

characterisation of the endosymbiont diversity in giant clams found that they primarily associate with Symbiodiniaceae from the genera *Symbiodinium*, *Cladocopium* and *Durusdinium*, and individuals can maintain multiple endosymbiont genera and species simultaneously. This diversity of Symbiodiniaceae in a giant clam may also affect the host's attributes, such as growth rate, reproduction and photosynthetic efficiency.

As the giant clam grows larger in body size, the reliance on phototrophy increases to meet its energy demands. In some species, such as *Tridacna gigas*, *Tridacna mbalavuana* and *Tridacna derasa*, phototrophy alone may provide most, if not all, of the carbon requirements. The acquisition of nutrients from a combination of phototropy and heterotropy has been said to explain why giant clams have rapid growth rates and, thus, likely allowing them to achieve their massive body sizes compared to other bivalves. But this heavy reliance on symbiosis to obtain nutrition, especially with increasing body size, could make giant clams more sensitive to elevated ocean warming caused by global climate change. Nevertheless, studies suggest that the composition and abundance of endosymbionts may shift within an individual clam, often due to coping with changes to the environmental conditions. Also, considering the diverse ecological attributes of each Symbiodiniaceae genus, the high diversity of endosymbiont species hosted by giant clams could further raise their resilience during environmental disturbances.

The Ecological Characteristics of Symbiodiniaceae algae found in Giant Clam Species

Symbiodiniaceae genus	Ecology of genus
Symbiodinium	Members of this genus are globally distributed, and most are adapted to living in shallow-water environments under high or variable light conditions.
Cladocopium	Members of this genus are known to associate with a broad range of host taxa such as cnidarians, molluscs and sponges; they are physiologically diverse, adapting to a wide range of temperatures and irradiances across habitats and spanning the intertidal to mesophotic zones.
Durusdinium	Members of this genus are known as extremophiles, with their diversity and distribution centred in the Indo-West Pacific; they are usually found in symbiosis with hosts living in stressful environments with significant diurnal or seasonal shifts in temperature or broad fluctuations in water turbidity, thus making them resistant to disassociation.

Source of information: LaJeunesse et al. (2018).

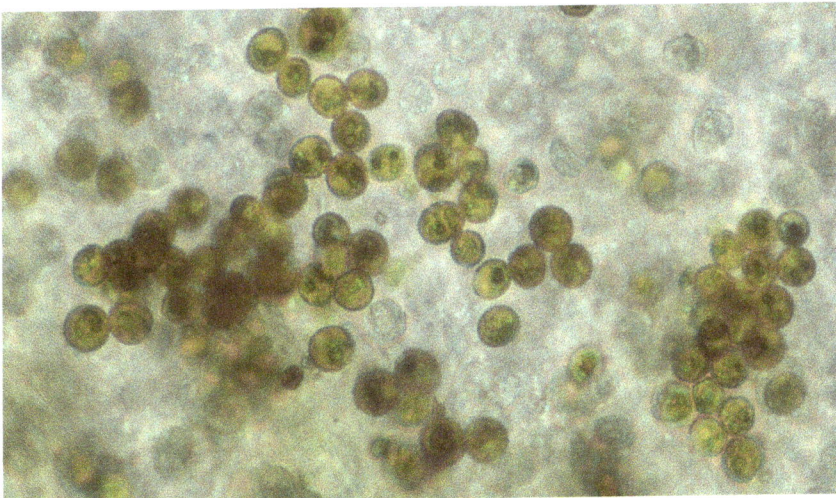

An image of the Symbiodiniaceae cells within the *Tridacna crocea* larvae. Establishing these microalgae cells in giant clam larvae is critical for long-term survival. ©Samuel Lee

Giant clams reproduce through broadcast spawning, releasing large amounts of eggs and sperm into the water column. Their gametes get into contact and fertilisation occurs externally.

Reproduction and Development

Mature giant clams are simultaneous hermaphrodites, where the same individual contains both male and female sex organs capable of producing both types of gametes: sperm and eggs. The gonads are not separated into male and female regions, but the spermatogenic and oogenic tissues occur in close proximity throughout the gonads. During spawning, the sperm and eggs are released as separate events presumably to prevent self-fertilisation. The sperm is released first, followed by the eggs that may only be released after more than an hour, although in most cases only sperm is released. Among the giant clam species, the fecundity highly varies, where the larger individuals or species can typically release up to hundreds of millions of eggs in a single spawning.

Upon the discharge of viable gametes through the excurrent siphon into the water column, fertilisation between sperm and eggs occurs externally. The fertilised embryo hatches into a swimming trochophore larva within the first 24 hours, and develops into a D-shaped veliger within the next 24 hours. The pelagic veliger drifts in the sea and filter feeds with its ciliated velum, and grows into a pediveliger over the next four to seven days. During these developmental stages, the larva acquires its endosymbiotic algae via ingestion from the environment independently, as these are not transmitted from parent to offspring. The endosymbionts will then manifest within the veliger's guts and selectively not be digested to establish in the siphonal mantle. There is little change in shell form between the veliger and pediveliger, where the latter later becomes an umbo-stage larva even though the umbo remains indistinct.

The competent pediveliger actively uses its foot to search and settle on suitable habitats. Then, it continues to metamorphose into a juvenile clam that attaches to the substratum by byssal threads. Settlement assay experiments have found that the competent giant clam larvae can be induced to settle and metamorphose more quickly in the presence of other conspecific larvae or the crustose coralline algae, a well-known inducer for larval settlement. After settlement, the increasingly sedentary juvenile clam can still move about by breaking the byssus and actively searching along the seabed on its extended foot for ideal refuges. The movement and aggregation behaviours in juvenile giant clams are hypothesised to serve several ecological roles, including defence against predation, physical stabilisation and facilitation of reproduction. Although members of giant clams have high reproductive outputs, the success in larvae survival and recruitment in the wild is considerably low. To further compound matters, early juvenile clams have slow initial growth rates, reaching shell lengths between 2 cm and 5 cm in their first year, which makes them highly vulnerable to predation. However, once the individual reaches its escape size, the giant clam is relatively predator-free throughout the rest of its life.

Life Cycle of a Giant Clam

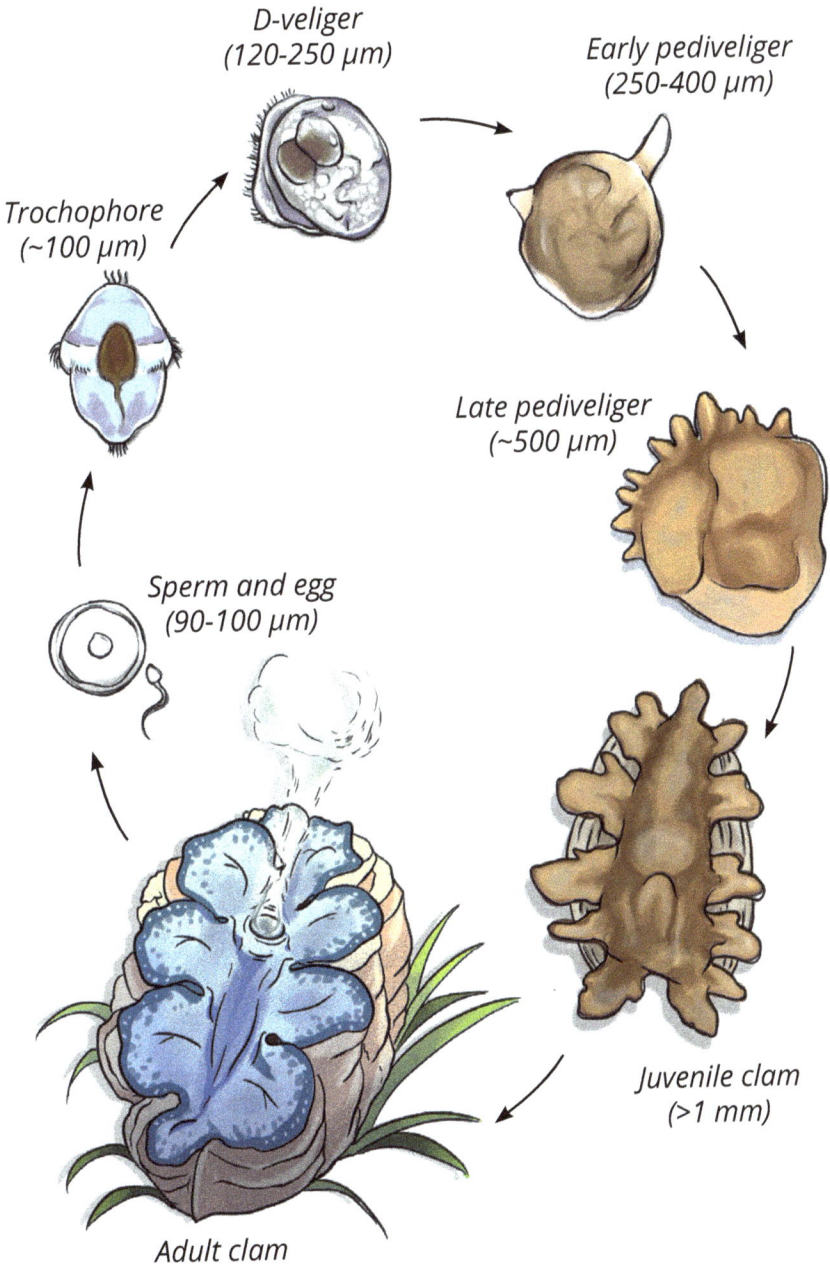

D-veliger
(120-250 μm)

Early pediveliger
(250-400 μm)

Trochophore
(~100 μm)

Late pediveliger
(~500 μm)

Sperm and egg
(90-100 μm)

Juvenile clam
(>1 mm)

Adult clam

Life Cycle of a Giant Clam. ©Adel Lee

BORING for Life - How *Tridacna crocea* Conquers a Rock!

Tridacna crocea is the smallest species of giant clams, which grows a shell up to 15 cm long. It is also the only species to thoroughly embed itself into coral rocks. After the young clams of this species settle on the seabed, they burrow into the coral reef substrata, thus effectively trapping themselves in their self-made boreholes for the rest of their lives. Only the brightly coloured outer mantle that extends beyond the borehole is in view, which allows the animal to capture sunlight for its endosymbionts' photosynthesis. When disturbed or threatened, giant clams, in general, can fully retract its outer mantle back into the shells to protect their vulnerable soft body parts. *Tridacna crocea* can further hide within the tight crevice, 'vanishing' from sight. The secret behind this burrowing behaviour in *Tridacna crocea* remained an enigma until recently.

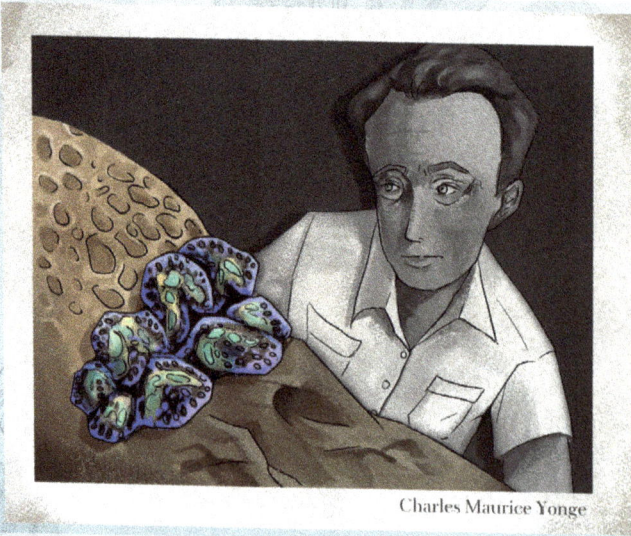

Charles Maurice Yonge

Sir Charles Maurice Yonge (1899-1986), a notable malacologist, was one of the earliest to ponder the mechanism of boring in *Tridacna crocea*. As the coral rock is made up of limestone subject to acid dissolution, Sir Yonge proposed that the clam has a boring organ that secretes acid to dissolve the substratum gradually. He then conducted pH experiments testing the seawater surrounding the clam but found no conclusive evidence for acid secretion. After refuting his own initial hypothesis of boring by acid secretion, Sir Yonge was still unable to pinpoint the mechanisms. His final conjecture was that *Tridacna crocea* bores entirely by mechanical grinding. He described that the animal grinds its way downwards by rocking back and forth along its length, gradually eroding the coral rock.

Richard Hill and his colleagues decided to re-examine this mystery by using newer tools to measure pH and consider the animal's anatomy and behaviour during boring. Upon removal of *Tridacna crocea* from its borehole, the whitish tissue that emerges from the byssal orifice on the underside of the clam is probably the "boring organ". This organ is known as the pedal mantle, which is greatly enlarged only in this species and not for other giant clam species. Considering the position of the pedal mantle, the researchers realised that direct contact of this boring organ to the surface is needed to detect pH effectively. So, the relaxed clams were carefully placed onto pH-sensitive foils to measure the surface pH of this pedal mantle for comparison to the surrounding seawater.

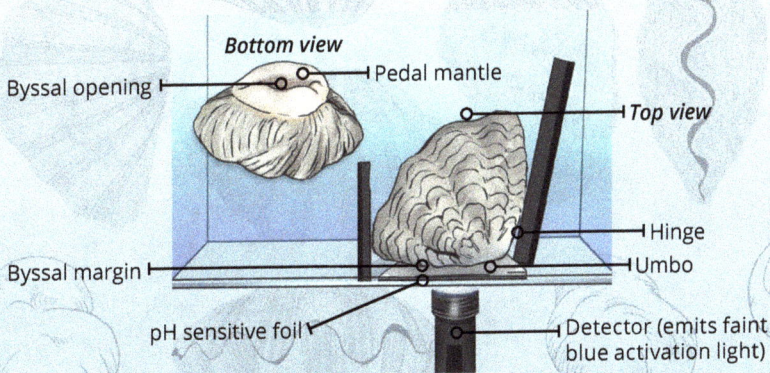

Bottom view
Byssal opening — Pedal mantle
Top view
Byssal margin — Hinge — Umbo
pH sensitive foil — Detector (emits faint blue activation light)

The experiment was a success! The pedal mantle's surface pH was significantly lower than the surrounding seawater pH, where the minimum pH measured at the pedal mantle's surface was between 4.65 and 5.36. These pH values were approximately two units lower than the surrounding seawater pH at 8.2! Further analyses also found an abundance of a particular enzyme called vacuolar-type H^+-ATPase that facilitates the release of H^+ ions from the pedal mantle. The latter lowers pH to acidify contact surfaces. Now we know how the *Tridacna crocea* conquers the coral rock and is boring in the very best sense of the word.

4. Ecological Roles in Coral Reefs

More than just a pretty mantle, the giant clams can provide essential ecological services and influence their surrounding environment. They are critical foundation species in the highly biodiverse coral reef ecosystem and efficient ecosystem engineers fulfilling multiple roles as reef builders, shelters, microalgae reservoirs, water filters, and even as "meals" for reef inhabitants.

A fluted giant clam (*Tridacna squamosa*) colonised by a diverse array of epibionts such as *Sinularia* soft coral (background), crustose coralline algae, ascidians and sponges (foreground). ©Mei Lin Neo

A marine environment is dangerous for many of the smaller reef fishes and invertebrates, and finding suitable protection is crucial to their survival. Their solution is to find a 'bodyguard' as burly as a giant clam! The presence of

giant clams can increase the topographic relief of the seabed and serve as functional shelters and nurseries for reef fishes. This increase in reef complexity is said to positively affect juvenile fish recruitment and settlement by providing refuge from predation. The highly sculptured shell valves also serve as micro-habitats for mobile organisms such as fish and snails and sessile organisms such as corals, sponges, ascidians and tubeworms. Alive or dead, these giant clam shell valves present extra surfaces for reef organisms to colonise, especially when there are limited areas of exposed surfaces on the reefs. Although not seen often, commensal shrimps and pea crabs seek long-term sanctuary inside the mantle cavities of giant clams. If you're lucky, you may catch a glimpse of them moving around the gills!

The commensal Pontoniinid shrimp (*Anchistus* sp.) found resting on the siphonal mantle of a fluted giant clam (*Tridacna squamosa*). ©Mei Lin Neo

Of Shrimps and Pea Crabs - The Elusive Neighbours Living with Giant Clams

Bivalves often host a wide diversity of commensal animals and provide the latter with refuge and food. However, little is known of these commensal animals living inside the giant clams as they are often hidden within the mantle cavity and ctenidia. The discovery of these elusive animals is usually opportunistic when the host giant clams are collected for research. In addition, these collected samples of commensals typically exhibit little signs of damage, which suggests that life within the giant clam is sheltered.

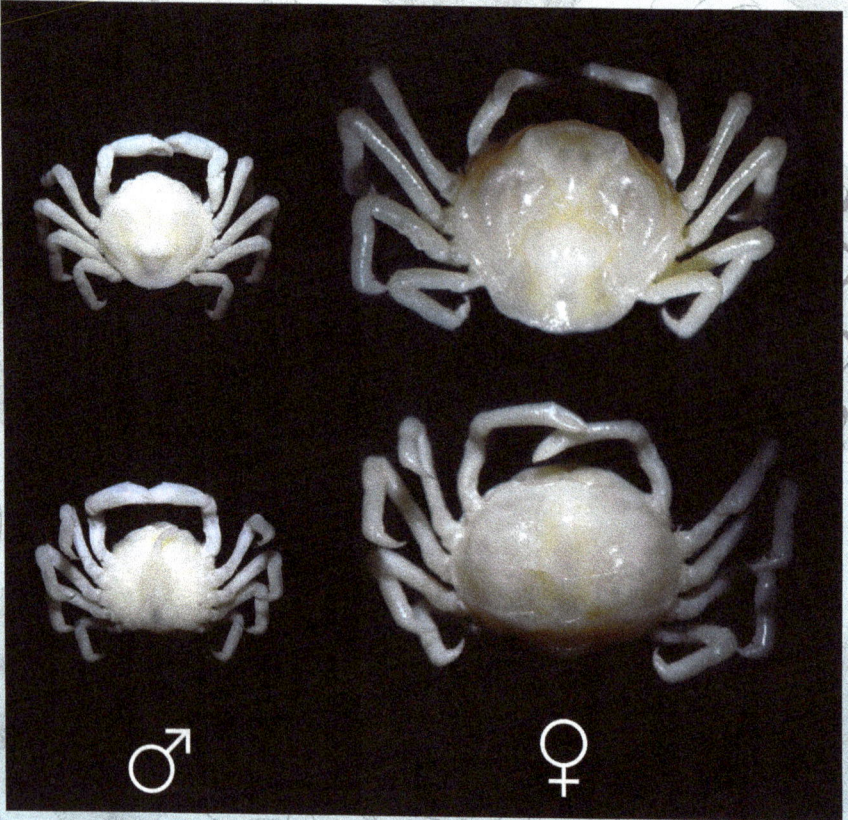

Among the Pinnotherid pea crabs associated with host giant clam, *Xanthasia murigera* (male on the left and female on the right) is the most widespread species across Indo-Pacific and has been found in five species of giant clams (*Hippopus hippopus*, *Tridacna crocea*, *Tridacna gigas*, *Tridacna maxima*, *Tridacna squamosa*). ©Mei Lin Neo

Pearl of Knowledge

So far, several species of commensal shrimps (Pontoniinids) and pea crabs (Pinnotherids) have been found living within the mantle cavities of various giant clam species. These commensal species typically have hooked walking legs that allow them to hold onto the ctenidia firmly. This movement will enable them to gain access to food aggregated by the host and anchor themselves against the strong currents generated by the ctenidia to avoid expulsion. It is hypothesised that a single long-lived giant clam can host many generations of commensals!

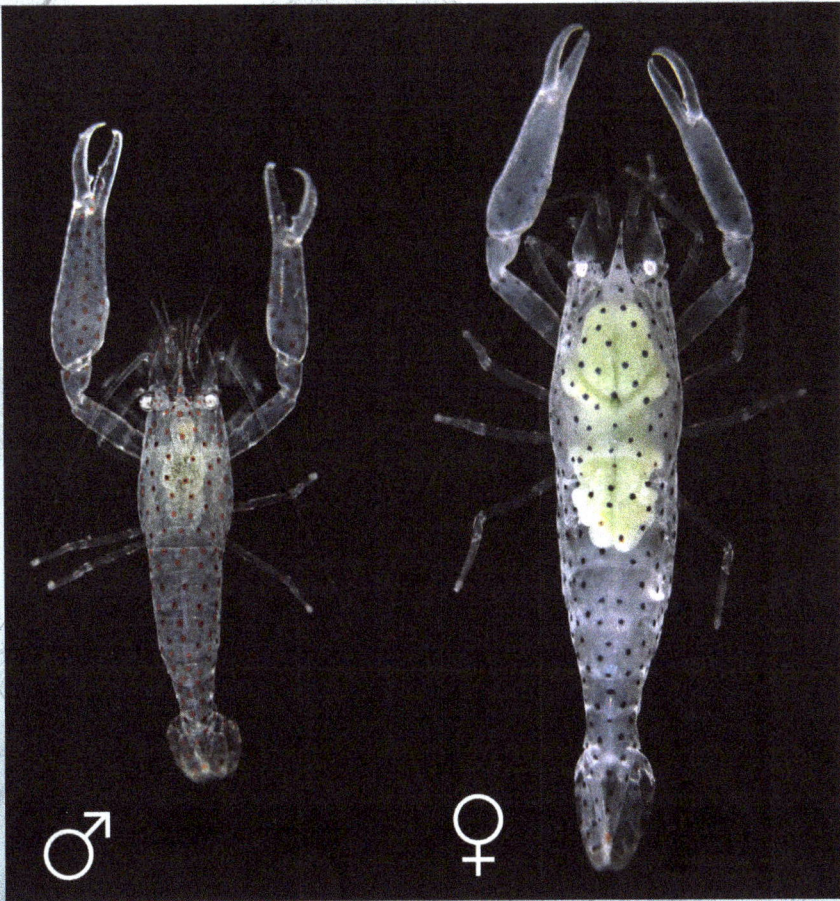

A possible mating pair of *Anchistus miersi* shrimps (male on the left and female on the right) collected from a fluted giant clam (*Tridacna squamosa*) in Singapore. ©Mei Lin Neo

A giant clam, is after all, an edible shellfish in the sea. A population of *Tridacna gigas* clams has the potential to amass up to 29,000 kilograms (64,000 pounds) of flesh per hectare annually, making them one of the top producers in the coral reef ecosystem. With this amount of seafood available, giant clams are attractive prey items, especially the juveniles, where nearly 75 species of reef animals have been observed to feast on them, including the pufferfish, crusher crabs, predatory snails (families Pyramidellidae and Cymatiidae), and even flatworms! Their highly nutritious faeces, containing mucus and proteins, are also a nourishing food source for other sea creatures such as the black damselfish (*Neoglyphidodon melas*). Besides, their faeces contain photosynthetically viable endosymbionts that may become available to other symbiotic species, including young giant clams.

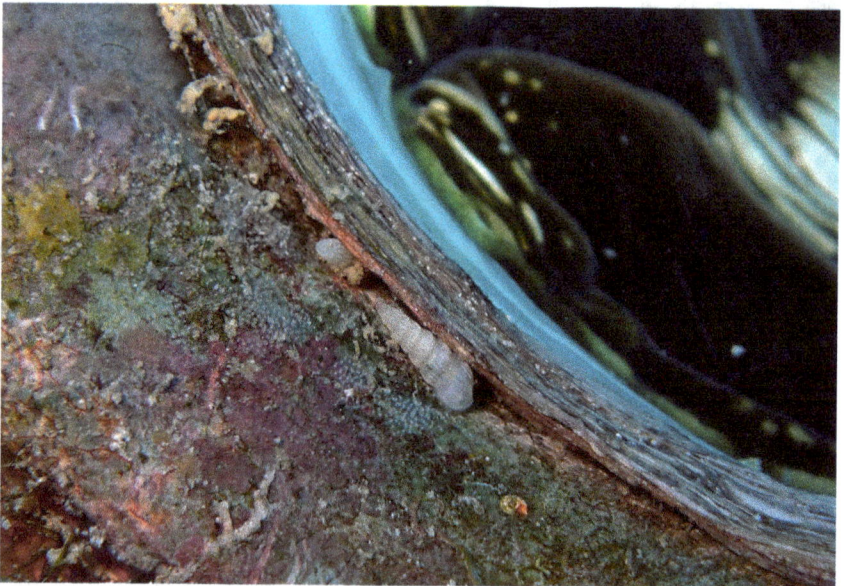

Pyramidellidae or pyramid shells are tiny ectoparasitic marine snails that often cause mass mortalities among cultured juvenile giant clams. Occasionally, these ectoparasites may be seen on wild giant clams such as the ones in this photo. These ectoparasites would hide within crevices during the day and feed at night. Mature snails would lay down egg masses (adjacent to snails) on the shell surfaces. ©Mei Lin Neo

Although giant clams are seldom perceived as major carbonate contributors to the reef framework, these animals can 'build' reefs with their heavily calcified aragonitic shells. For instance, the highly dense *Tridacna maxima* populations (i.e., around 900,000 individuals) in Tatakoto atoll of French Polynesia could produce as much as 100,000 kilograms (220,000 pounds) of shell material per hectare annually! This extraordinary abundance and dominance of *Tridacna maxima* clams is well-known in the atolls of French Polynesia as they form small islands known as '*mapiko*', which means the natural accumulation of dead and live clam shells in the local Tuamotu language. These large-bodied animals may also modify the seabed by modulating water flow around them or adding habitat complexity in areas where they occupy. Even after the giant clam dies, the heavy shells that remain behind continue influencing water flow in the environment. In addition, some giant clam species, such as the *Tridacna crocea*, are known to efficiently 'destroy' coral rocks. The boring clams embed themselves in dead or living coral patches and, in the process, excavate carbonate from the reefs. Hence, this biological erosion clears out dead coral heads, making room for new corals to grow.

Given their numerous ecological roles in the marine environment, giant clams have been lauded as ecosystem engineers that can directly or indirectly influence the habitats. In addition, they could serve as bioindicators in monitoring reef health as these filter-feeding sentinels efficiently bio-accumulate pollutants in their tissues and organs that can be measured as proxies for pollution levels. All in all, only a few marine organisms can match up to the giant clam's ecological contributions in the coral reef ecosystem. Hence, it is crucial to protect and manage giant clams in the coastal habitats they inhabit, as their persistence brings about significant ecological benefits to other marine organisms and ensures the continued functioning of habitats.

©Roger Dolorosa

Giant clams have a giant impact on coral reefs. These multi-tasking clams are reef builders, food factories, shelters for corals, sponges, shrimps and crabs, and water filters - all rolled into one! In a nutshell, giant clams play a major contributing role as residents of their own reef home, and just having them around keeps the reef healthy.

Clam-O-Meter - Giant Clams as Pollution Indicators

"Clearly, tridacnid clams show considerable promise as indicators of trace metal pollution in tropical waters. ~ Denton & Winsor, 1986"

Heavy Metals

Trace heavy metals are present in coastal environments due to natural and anthropogenic processes. Thus, it is unsurprising that studies have found that giant clams accumulated trace metal elements such as copper, zinc, cadmium, cobalt, manganese, nickel and lead. These trace metals are usually found in the tissues of their organs, especially in the kidney, mantle and gills. A biomarker called metallothioneins (MTs), a group of metal-binding proteins found in various tissues, is used in detecting trace metals in aquatic animals. These MTs protect against heavy metal damages, where they can bind toxic metals for detoxification. In the giant clam, a form of MTs called MT-like proteins was found with similar protective functions. A study found that the increase in concentrations of MT-like proteins corresponded with the higher levels of total trace metals found in the kidneys of clams collected from the wild. This evidence shows that the giant clam can be a valuable bioindicator for inferring environmental levels of heavy metals. It is also essential to know how much trace metals are present in the giant clam tissues, especially in populations near the coast where they are taken as food.

Ciguatoxins

Ciguatoxins (CTXs) are a group of neurotoxins produced by dinoflagellates (a type of microalgae) of the *Gambierdiscus* genus that grow on and around coral reefs in tropical and subtropical waters. In the Pacific Ocean, the Indian Ocean and the Caribbean Sea, ciguatera fish poisoning (CFP) is the most common non-bacterial seafood poisoning that results from eating tropical coral reef fishes contaminated with CTXs. However, several Pacific Islands countries, including New Caledonia, French Polynesia and Vanuatu, reported atypical ciguatera-like incidents involving giant clams. Since they are efficient filter-feeders, they could bioaccumulate the *Gambierdiscus* phytoplankton and its neurotoxins in their bodies, thus serving as a novel pathway for human seafood poisoning via bivalve molluscs. Under experimental conditions, the giant clam *Tridacna maxima* was exposed to two strains of *Gambierdiscus* - a highly toxic strain versus a weakly toxic strain.

Interestingly, experimental clams exposed to the highly toxic strain had bioaccumulated CTXs at concentrations well above the safety limits recommended for human consumption. In contrast, those exposed to the weakly toxic strain were toxins-free. This suggests that, in nature, the risk of CTXs contamination in giant clams is viable in the presence of highly toxic *Gambierdiscus* blooms, confirming that these bivalves can be another pathway in ciguatera poisonings. This new evidence also highlights the need to monitor other marine invertebrates in affected areas for an accurate assessment of ciguatera risks to humans.

Microplastics

The ingestion of microplastics or MPs (broadly classified as particles <5 mm in size) in large marine mammals to the tiny planktonic larvae has been well-documented. These MPs may cause obstruction and injury to their digestive systems, starvation or malnutrition due to reduced food intake, and the potential translocation of MPs into the major circulatory system. Also, the persistence of plastics in the bodies could lead to leaching of hazardous chemicals into the surrounding tissues causing toxic effects. However, only few studies have reported the ingestion of MPs and its impact on giant clams. A giant clam would largely behave similarly to other filter-feeding clams when capturing MPs from the water column. Still, the uptake rate would differ across bivalve species as they may discriminate particles based on the size, shape and surface properties. Studies exposing polystyrene microbeads of different sizes and concentrations to two giant clam species (*Tridacna maxima* and *Tridacna crocea*) confirmed that they actively take up these MPs from the water column. Like most other bivalves, microbeads were predictably found in the clam's digestive tract and excretory faeces. One of the studies surprisingly found that shells were a major sink for MPs, where about 66% of microbeads had passively attached to *Tridacna maxima*'s outer shell valve surfaces. Another study found that the ingested MPs could negatively affect metabolic processes such as photosynthesis and nutrient transport in *Tridacna crocea*. By and large, the giant clam can be used as a bioindicator for MP loading in the coral reefs since they are shown to be effective sinks for MPs.

©Heok Hui Tan

5. Cultural and Socioeconomic Significance

In the years gone by, giant clams were notorious for their supposed man-eating abilities, which of course, is not true. Instead, they held great importance in societies and made up a significant part of fisheries across the Indo-Pacific for more than a millennium. Giant clams are valued because they offer food security and income, maintain traditional customs, yield socio-ecological benefits (i.e., improved reef health), and benefit coral reefs. Growing in popularity, the aquarium trade for giant clams is also increasingly significant over the past three decades. In some areas, giant clams are strongly featured in local folklore of coastal communities living in the region.

"A giant clam was harvested from the shallow waters near a village in Tabiteuea of the Republic of Kiribati. These animals are occasionally kept in villages as quasi-domestic animals that can feed an entire family in times of need. Their large shells are commonly used as pig troughs as well as baptismal altars in some of the churches." ©George Steinmetz

Meat as Food

Giant clam meat has been a significant component in people's diets in Oceania and Southeast Asia. Like other nutritious molluscs, the giant clam's mantle flesh and adductor muscle are good sources of protein, amino and fatty acids, with low fat content. The entire clam, except for its kidneys, can be eaten as a raw or cooked dish. Giant clams were formerly regarded as 'stormy weather food', the fresh staple that was easy to collect and have on hand when the weather was too stormy to go fishing offshore. While all species of giant clams have been collected for food, coastal populations in different geographic locations have shown specific preferences towards certain species based on their meat texture and availability of stocks.

A fisherman's feast of seafood that includes a platter of giant clam meat, where each major organ is separately plated for consumption. Kumejima, Okinawa, November 2009. ©Heok Hui Tan

● Mantle

● Adductor muscle

● Innards, gonads

Eaten by a Giant Clam

The 'man-eating' giant clam gained fame as the evil villain known to swallow a whole man with its jaws! Unfortunately, this cultural belief between the early 1920s and 1940s led to a misunderstanding of these gentle giants known for their enormous body sizes and weights. The US Navy Diving Manual was said to give detailed instructions on how to release oneself from the grasp of the clam by severing its adductor muscle that holds the shell valves together. In other accounts, they even mentioned the drowning of a pearl diver when a *Tridacna* shell clamped onto his leg!

Pearl of Knowledge

By the 1960s, when people started studying these large bivalves, they quickly realised their docile nature. And while the clam does close its valves in the presence of divers and snorkellers, they usually do it slowly, giving ample time for people to react and escape. Moreover, the outer mantle flesh in larger clam species is so meaty that it usually cannot fully clamp up. But, admittedly, the smaller clam species can still possibly grip onto your fingers if you're not careful enough!

Shell as Material

The use of giant clam shells as raw materials and decorative ornaments has had a long history worldwide. They were a well-traded commodity across the Near East (modern-day Middle East region), where engraved *Tridacna* shells were frequently uncovered from the archaeological sites in Iran, Jordan, Israel, Iraq, Libya, Egypt, as well as in Italy and Greece, where these shells do not occur naturally. Their shell valves were predominantly traded for decorative and functional purposes such as producing high-value products including jewellery and body ornaments to low-value uses, such as an aggregate in building materials. The domestic uses of giant clam shells are also wide-ranging among the coastal communities, including as traditional ceremonial objects, ornaments, and tool use such as adzes and drinking containers for livestock. In recent times, the shells have been used in the shell craft industry as carving materials to produce high-value decorative ornaments that resemble white ivory; hence these items are often dubbed the "Jade of the Sea".

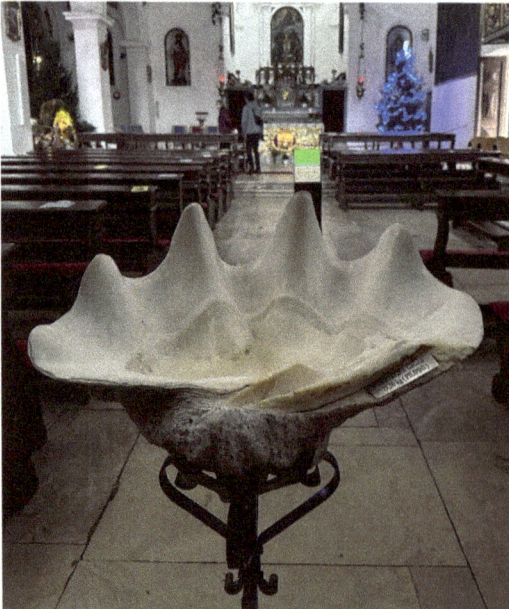

In the European Catholic Churches, giant clam shells, particularly of *Tridacna gigas*, are often used as baptismal fonts (vessels containing holy water). Originally from the Philippines, this specimen was found in a church in Piran, Slovenia. ©Samuel Lee

Live Ornamental Trade

For their brightly coloured and patterned mantles, the giant clams were keenly brought into the aquarium trade around the early 1990s. Despite a genuine interest from the aquarium retailers to sell the giant clams, many had expressed concerns over the need for more knowledge on aquarium care, transportation, and the availability of stocks to sustain the aquarium trade in the long term. With time, the accumulation of knowledge boosts the confidence of both retailers and hobbyists in keeping giant clams as aquarium pets. In 2001 alone, more than 100,000 live giant clams were internationally traded for the aquarium industry, and that number has since been growing rapidly. Between 2007 and 2016, the major countries exporting live individuals of *Hippopus* or *Tridacna* species for the wildlife trade were Viet Nam, French Polynesia, Cambodia and Fiji.

Giant clam currently sold in the aquarium trade are predominantly aquacultured, with only a few countries having permissions to sell wild-collected stocks. These brightly coloured individuals can typically fetch up to a couple hundred dollars, while the rarer species can cost as much as a few thousand dollars. ©Mei Lin Neo

Giant Clam Folklore Around the World

Europe
Due to their massive sizes, the *Tridacna gigas* shells gained fame among European explorers. These shells were brought back to Europe and revered by collectors and kings. These shells later made their way into Christian churches and rituals as baptismal fonts.

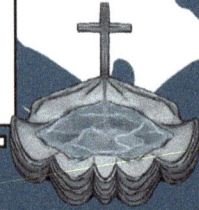

China
In the Buddhist scriptures from the Han Dynasty, the giant clam (locally known as 硨磲, *che-qu*) is mentioned as one of the seven treasures of nature. Wearing ornaments made from giant clams is believed to calm the spirit, remove negative energy, and bring good luck and prosperity.

Near East
Numerous engraved *Tridacna* shells were uncovered from archaeological sites in the countries of the Near East (modern-day Middle East region). The shells were most likely used as containers for cosmetics, jewellery or foodstuff.

Palau
The Palauans' creation story began with a giant clam called to life in an empty sea. The primordial clam grew bigger until it sired Latmikaik, the Palauan Goddess of the Sea. She was also the mother of human children, who birthed them with the help of storms and ocean currents. In addition, shells were often used as ceremonial containers for ancestral skulls or ritual washing vessels.

Solomon Islands

In local churches, the *Tridacna gigas* shells are used as containers or covers for ancestral skulls and as ritual washing basins for priests during sacrifices. In addition, shell rings, made from fossilised shells, were used as an exchange currency, ceremonial objects and tribal heirlooms for leaders.

Ryukyu Islands

Giant clams are highly beloved by Okinawan islanders as they are an important heritage food in traditional Okinawan food. Clam shells are also used as ceremonial burial objects or amulets (魔除け, *mayoke*) as they are believed to have magical powers to capture the dead spirits, prevent curses and ward off evil. The locals of Miyako and Yaeyama Islands share the belief that before humans, the first objects created by the gods were clams.

Sulu Sea, Philippines

Native islanders feed their young ones with *Tridacna squamosa* to strengthen their teeth. Giant clam meat was also said to be a good source of iodine and an alternative food item for fish.

Sabah, East Malaysia

The Rungus and Bajau ethnic groups still use *Tridacna* shells as raw materials for traditional shell artefact production. Shells are made into body ornaments such as necklaces and bangles, often used as accessories in traditional costumes.

Indonesia

Natives of some seafaring communities in Indonesia believed that eating the adductor muscle meat of giant clams can increase men's vitality, therefore fetching prices up to USD$20 per kg. *Tridacna crocea* (locally known as *Kima Lubang*) is also highly sought after as it is a popular belief among housewives and elders that it can help increase milk production in nursing mothers.

6. Threats and Challenges

Giant clams are an all-round coastal resource to the local fishing communities and commercial markets throughout their geographic ranges. Because of their shallow distribution, conspicuous appearance and sessile nature, these large shellfish make easy targets for harvesting by hand collection. SCUBA and improvised diving apparatus such as hookah gear (a simple surface air feed) are also used to reach individuals living in deeper waters. The entire animal is usually collected, where the fleshy meat is removed from the shells using knives, wooden sticks or metal stakes. Across the Indo-Pacific, most of the coastal communities consider giant clam meat a delicacy to be consumed only during special occasions; thus, they are only collected opportunistically during reef gleaning, free-diving and non-specific fishing.

On the other hand, for some communities, the giant clam is a primary target of fishing trips in areas where their densities are still high. Nearly all of the giant clam species have been and continue to be exploited for their meat as food, fish bait or animal feed, their shells sold to the curio trade, or the live individuals to be exported for the aquarium trade.

Over the years, many of the wild giant clam populations across the Indo-Pacific have since declined drastically as a result of previous overfishing exploits either for commercial markets or subsistence use. Before the 1980s, the high commercial demand for their adductor muscles in the Asian markets had eventually led to widespread illegal poaching by long-range foreign vessels in the Pacific Ocean. Reef surveys have found that the abundance of giant clams, especially the larger species such as *Tridacna gigas*, *Tridacna derasa* and *Tridacna squamosa*, had plummeted to extremely low densities after bouts of intensive exploitations. Furthermore, the subsistence harvesting of giant clams could threaten the remaining stocks if not adequately managed. Today, the exploitation of wild giant clams for commercial trade has been banned in most countries, but poaching continues to threaten the remaining wild populations. Specifically, the coastal resource authorities of various countries, including Australia, Cambodia and the Philippines, have reported increased fishing boats poaching for giant clam shells illegally between 2015 and 2020. The heightened demand for these shells has been attributed to the flourishing Asian shell craft industry, where the large shells are typically carved into sculptures, and medium shells are processed into jewellery and religious prayer beads.

Giant clams are also increasingly threatened by other marine anthropogenic activities that can affect their fitness and survival. As the coastal habitats throughout Indo-Pacific are growingly encroached upon by the expanding human populace, these environments gradually become degraded due to various stressors, including coastal urbanisation, land reclamation, sediment pollution, heavy metal contamination and tourism. While the impacts on giant clams vary depending on the type of anthropogenic pressures, many studies have confirmed the adverse effects on their fitness, such as reduced survival, growth and photosynthetic performance, altered behavioural responses and increased respiratory demands to cope with these stresses. Furthermore, these mega bivalves are likely highly susceptible to human-induced pressures, as a review study led by Simon Van Wynsberge and colleagues had predicted that the densities of *Tridacna maxima* tended to decrease when there was an increase in human activities. Human populations and their activities are potentially significant drivers behind the extinction of giant clam species.

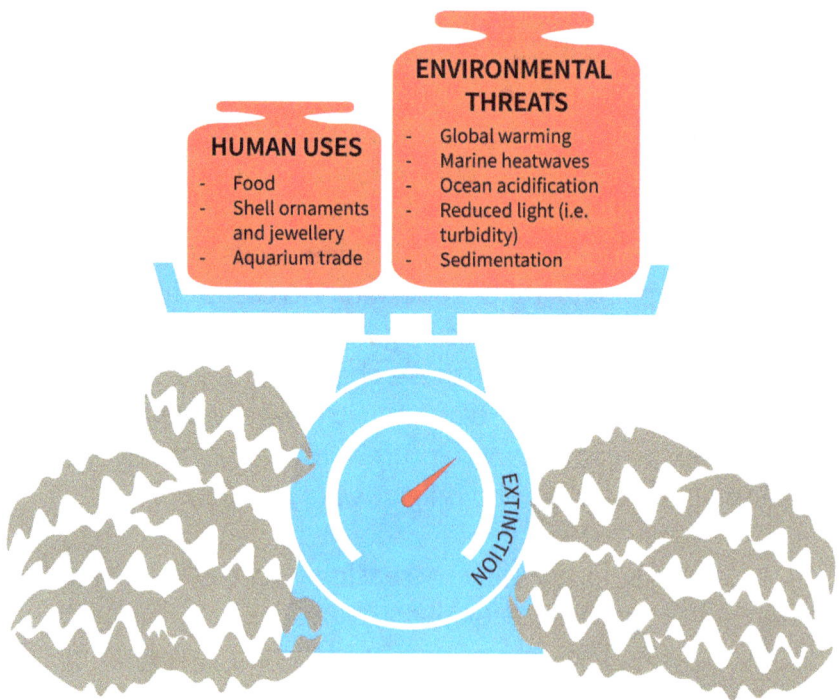

HUMAN USES
- Food
- Shell ornaments and jewellery
- Aquarium trade

ENVIRONMENTAL THREATS
- Global warming
- Marine heatwaves
- Ocean acidification
- Reduced light (i.e. turbidity)
- Sedimentation

EXTINCTION

Giant clam populations across its geographic range continue to face increasing pressures from human overexploitation, agriculture, urbanisation and global change, which may push them towards extinction.

Among the environmental threats, the current climate crisis driven by global warming may have a hand in threatening the survival of giant clams. Global warming is the long-term warming of the Earth's overall temperatures primarily due to burning fossil fuels that produce heat-trapping greenhouse gases. As a result, the oceans are absorbing more heat, which causes an increase in sea surface temperatures and rising sea levels. These extreme conditions, either in elevated sea surface temperatures or ultraviolet irradiation, can lead to undesirable effects on the giant clams, including poorer growth, higher mortality, and lower reproductive performance. Giant clams are particularly vulnerable to warming oceans because of the animals' unique symbiotic relationship with the Symbiodiniaceae microalgae. Under elevated temperatures and high light intensities, these shellfish could experience bleaching, which causes the expulsion of their photosynthetic endosymbionts, which turns their fleshy tissues white. Even though they are at considerably higher risk against global warming, information on the bleaching incidences among giant clams remains scarce. Of the available records, they generally described a large number of affected clams, as well as a consequential high rate of mortality among the bleached individuals. As the planet is predicted to continue warming due to the elevated levels of anthropogenic greenhouse gas emissions, these photosymbiotic clams inhabiting coral reefs are expected to be impacted by rising seawater temperatures, resulting in more frequent mass clam bleaching and mortality events.

Reports of Giant Clam Bleaching across the Indo-Pacific

Year of bleaching	Where and what happened?
1997-1998	*Great Barrier Reef (GBR), Australia:* Seawater temperatures had increased from 22°C to 32°C within several weeks, accompanied by heavy rainfall that lowered the salinity to 18 ppt. Extensive bleaching of cultured clams and wild stocks was observed in the central region of GBR. More than 8,000 out of the 9,000 *Tridacna gigas* individuals were bleached.
1998	*Bolinao Ocean Nurseries, Philippines:* Seawater temperatures reached 34.9°C and 34.1°C in June and July 1998, respectively. Mortalities were observed for all species cultured in the nurseries, with the worst performer being *Tridacna gigas*.
1998	*Southern Seychelles, Western Indian Ocean:* Sea surface temperatures did not fall below 30°C for the entire period from February to April 1998. Bleaching was widespread in other coral organisms, including the giant clams (*Tridacna* spp.).
1998	*Takapoto Atoll, French Polynesia:* Seawater temperatures rose above 30°C for over five months from December 1997 through April 1998, peaking at 31.8°C in mid-March 1998. Extensive bleaching of *Tridacna maxima* in Takapoto atoll was observed, with some partial recovery in May 1998. However, the abundance of giant clams has decreased by more than 80% since 1993.
1998	*Western Caroline Islands, Palau:* Local sea surface temperature was 31°C for at least 30 days during the late summer of 1998. Early qualitative observations found that many other coral reef organisms, including giant clams (*Tridacna gigas*), were bleached.
1998, 2009/2010, 2016	*Mauritius:* The rapid decline of giant clam numbers in Mauritian waters was attributed to the key coral bleaching events in 1998, 2009/2010 and 2016. However, studies are needed to assess the recovery rates of the giant clam populations following these bleaching events.

Year of bleaching	Where and what happened?
1998, 2010, 2016	*Lakshadweep Archipelago:* The increased frequency of seawater temperature anomalies has caused a gradual decline in giant clam numbers between 2004 and 2017. A mean summer sea surface temperature of >30°C triggered bleaching in *Tridacna maxima*, although these bleaching events did not cause instantaneous mortality of adult clams.
2009-2010	*Mannai Island, Thailand:* Seawater temperatures peaked at ~32.5°C in May 2010, with prolonged temperatures of more than 30°C from February to September 2010. Extensive bleaching of wild clams (*Tridacna crocea* and *Tridacna squamosa*) was observed, with almost 60% mortality after the event.
2010	*Koh Racha Yai, Andaman Sea and Koh Tao, Gulf of Thailand:* A thermal anomaly was observed in the Andaman Sea and the Gulf of Thailand from April to June 2010. The highest measured temperature in the upper Gulf of Thailand was 33.9°C. Bleaching of giant clams was observed, particularly affecting those in smaller size classes.
2015-2016	*Reao, Tuamotu Atolls, French Polynesia:* Seawater temperatures were above 30°C and up to 38°C in March 2016 in some shallow and enclosed parts of the lagoon in Reao. Surveys of *Tridacna maxima* in the Reao lagoon found that 76.7% of wild clams and 90% of cultured clams were bleached. Post-bleaching observations found that bleaching resulted in an 18-50% mortality rate in cultured clams, with the remaining surviving clams recovering their endosymbionts and associated colours.
2015-2016	*Tongareva Atoll, Northern Cook Islands:* For prolonged periods, the lagoon temperatures reached 38°C, and the nearshore sea surface temperature was 33-34°C. Observations suggest that 95% of *Tridacna maxima* had died due to prolonged elevated seawater temperatures.

Giant clams exhibited partial bleaching (*Tridacna maxima* on top and *Tridacna crocea* on bottom) at Koh Racha Yai, Phuket, Thailand, in 2010's global bleaching event. ©Mei Lin Neo

Another direct consequence of the increasing carbon dioxide levels in Earth's atmosphere is the gradual acidification of oceans. Ocean acidification is when the oceans absorb atmospheric carbon dioxide, reacting with seawater and lowering pH. This process causes a decline in the saturation state of seawater with respect to calcium carbonate polymorphs such as aragonite, which is the main form of calcium carbonate in giant clam shells. The result is a change in the water chemistry such that fewer carbonate ions, the primary building blocks for shells and skeletons, become available for uptake by the giant clams that produce large and heavily calcified shells. Experimental evidence has found that ocean acidification alone can reduce the survival rates and shell growth (dissolution of shells) in giant clams. In addition, when ocean acidification interacts with other abiotic stressors such as temperature, salinity and light availability, their effects on giant clams may differ. More studies are required to ascertain the magnitude of ocean acidification impacts on these marine megafaunas and focus on their early life-history stages, as small marine larvae are especially vulnerable to increased acidity.

The impacts outlined above can directly or indirectly reduce densities of populations across their ranges in the wild, which will have severe repercussions on the giant clams' ability to reproduce successfully. A key characteristic of giant clam reproduction is the dependence on synchronised spawning with other individuals, where the aggregation of mature adults is necessary to encourage reproduction. Giant clam populations are, therefore, susceptible to stock depletion as the sparse adult populations result in lowered (or zero) fertilisation success rates and, consequently, the reduced recruitment of juvenile clams. To compound matters, as stocks become more scarce, the harvesting size tends to decrease, meaning that individuals may be harvested even before reaching reproductive viability, which further affects the availability of mates and limits fertilisation rates. Therefore, losing viable spawning individuals could lead to functional extinction and the eventual collapse of entire populations. On a lighter note, the natural recovery of wild populations is possible as the planktonic larvae can travel from reef to reef via prevailing currents. Still, this process may take decades before populations return to their pristine states.

Cryptic diversity also poses another challenge for the managing and conserving giant clam species. Cryptic species usually consist of two or more distinct species that have been classified as a single species as they are more-or-less indistinguishable by morphology but are genetically divergent and incapable of interbreeding successfully. Considering the high variability among species' morphology, the giant clams are good candidates for crypticity. From the late 2000s onwards, the use of genetic data in biodiversity surveys has led to the rediscovery of several giant clam species, specifically *Tridacna squamosina*, *Tridacna noae*, *Tridacna rosewateri* and *Tridacna elongatissima*. Due to overlapping morphological similarities, these cryptic species were previously lumped with the more common species, *Tridacna squamosa* and *Tridacna maxima*. When species identities become confused, it can have implications for giant clam fisheries and their regulations, such as the potential for misidentification and misreporting. Furthermore, lacking knowledge of these cryptic species makes it difficult to establish their exact distributions in the environment. As we advance, the integrated use of morphological and molecular approaches towards species identification is earnestly needed to implement appropriate conservation measures and policies for the giant clams.

Jade of the Sea - The Rise of the Giant Clam Shell Craft Industry in China

An intricate elephant carving that preserved the overall fan-shaped shell outline, which is typically a diagnostic feature of a giant clam. ©Mei Lin Neo

In China, the giant clam has been highly regarded as one of the national treasures. Their shells, in particular, are deemed as a valuable organic gem (also known as 天然有机宝石 in Chinese) that take years to cultivate and produce their purest white shells. In ancient China, their shells were crafted into beads and often used as ornaments on the headwear of imperial servants during the Ming Dynasty and as prayer beads for the higher-ranked Tibetan Buddhist monks. Also touted as one of the seven treasures in Buddhism, the accessories made from giant clam shells are believed to have medical powers that could help wearers improve health, enhance wisdom and ward off evil. Moreover, giant clam shells are greatly desired by collectors as they are rare organic gems from the sea that exhibit unique growth lines, which make them difficult to counterfeit. This means that all handicraft pieces are one of a kind in the world, thus prompting their higher-than-average market prices. As these local beliefs are still prevalent, the manufacture and commercialisation of giant clam shell handicrafts have significantly grown in China.

When giant clam shells were first sold before the 1980s, they lacked appeal and were considered low-value items commercially. However, this changed in the early 1990s when a Taiwanese businessman opened the first shell-crafting factory in Tanmen Village on Hainan Island. It is widely known that a primary source of giant clam shells for the carving industry is the dead and abandoned shells found in the coral reefs of the South China Sea. For decades, the South China Sea has been a traditional fishing ground for the Chinese, occasionally harvesting giant clams for their meat while discarding their shells. Although the heavy and thick shells were previously considered of no great value and left behind only to be buried by the waves over the years, they are, in fact, suitable materials for carving. Therefore, the Chinese artisans, who primarily began their trade with jade, also exploited the clam shells as new carving materials. Remarkably, after processing, these shells give off a translucent appearance resembling ivory when carved, therefore fetching high prices among local consumers. These shells also come in various colours, such as red, purple, yellow, brown and white, and the blood-red colour variation is favourably valued due to its rarity.

From 2011 to 2016, the lucrative carving industry flourished tremendously due to a surge in the popularity and demand for handicrafts made from giant clams. For years, the giant clam shells formed the backbone of the local economy in Tanmen, which was once a quiet fishing town. Upon entering the village, it was clear that these large shellfish had permeated residents' livelihoods, and many spoke highly of their handicrafts. As a result, many traditional fishermen had converted to 'clam fishing', where they dedicatedly collect giant clam shells to supply the industry as source materials. As a result, the widespread mining for these fossilised giant clam shells has devastated large areas of coral reefs and seabed in the South China Sea. In addition, many of the jade carvers from mainland China turned to these shells as an alternative medium. They moved their families to Tanmen to profit from the booming shell carving industry. At the peak of this shell craft industry, there were nearly 460 retailers, compared to 15 in 2012, which supported around 100,000 people.

The giant clam shell craft industry abruptly stopped in January 2017 when the Hainan provincial government enforced a provincial ban on the environmentally destructive business. Many

A variety of handicrafts were made using giant clam shells. Top left photo: The top necklace consists of blood red-coloured beads and a bicolour pendant derived from giant clam shells. Top right photo: The top row of bracelets is made from white clam shells with hints of yellow streaks compared to the bottom row of bracelets made from the red corals. Bottom photo: Large giant clam shells are often used in carving symbolic icons such as the Eighteen Arhat (十八羅漢). ©Mei Lin Neo

storefronts were forced out of business, while any fishing vessels entering the port carrying clam shells were impounded, and there were routine confiscations of large shells from processing factories. While these measures appeared to have slowed businesses in Tanmen, the appeal of giant clams remained strong. The protected status of giant clams and the ban in Hainan became a selling point, as the scarcity of giant clam handicrafts inflated their value. Local conservationists who visited Tanmen in 2019 revealed that under the innocuous guise of selling legal products, these shops continue to covertly sell handicrafts made from endangered marine life such as giant clams, hard corals and sea turtles in secret backrooms. Other traders resorted to selling their stock pieces under the table through backdoor shipping channels or social media platforms to move their products off the island. Elsewhere in Hainan Province and other provinces in South China, shops openly sell giant clam shell products believed to have originated from Tanmen, considering these areas outside of Hainan are unaffected by the regulations. Various shop owners have claimed that their shells would not run out anytime soon, withholding details on their supply chain and the whereabouts of these stockpiles. For now, the Tanmen locals remain confident that the giant clam shell craft industry is here to stay, as they operate in the shadows waiting for the tides to change.

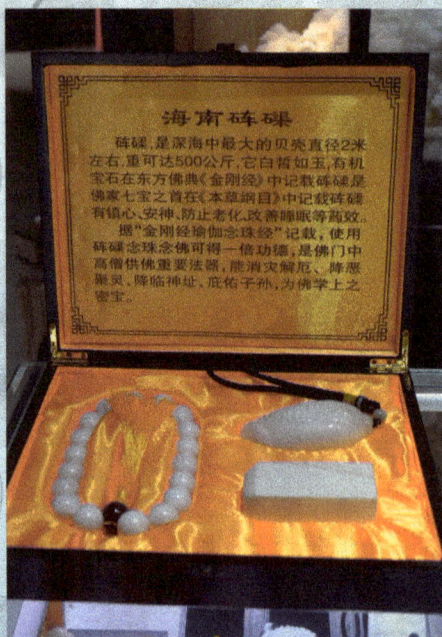

"Hainan Giant Clams": Ornaments made from giant clam shells are believed to bring benefits to wearers such as calming their mind, preventing aging, and improving sleep quality. According to the Diamond Sutra (金刚经), reciting the Buddhist scriptures using prayer beads made from *Tridacna* shells can double one's merits. These *Tridacna* beads are an important artifact that is believed to ward off disasters, conquer evil, and protect descendants. ©Mei Lin Neo

In Tanmen Village, large stockpiles of giant clam shells of *Hippopus* and *Tridacna* species were a common sight in large processing factories before the provincial ban by the Hainan local government. ©Mei Lin Neo

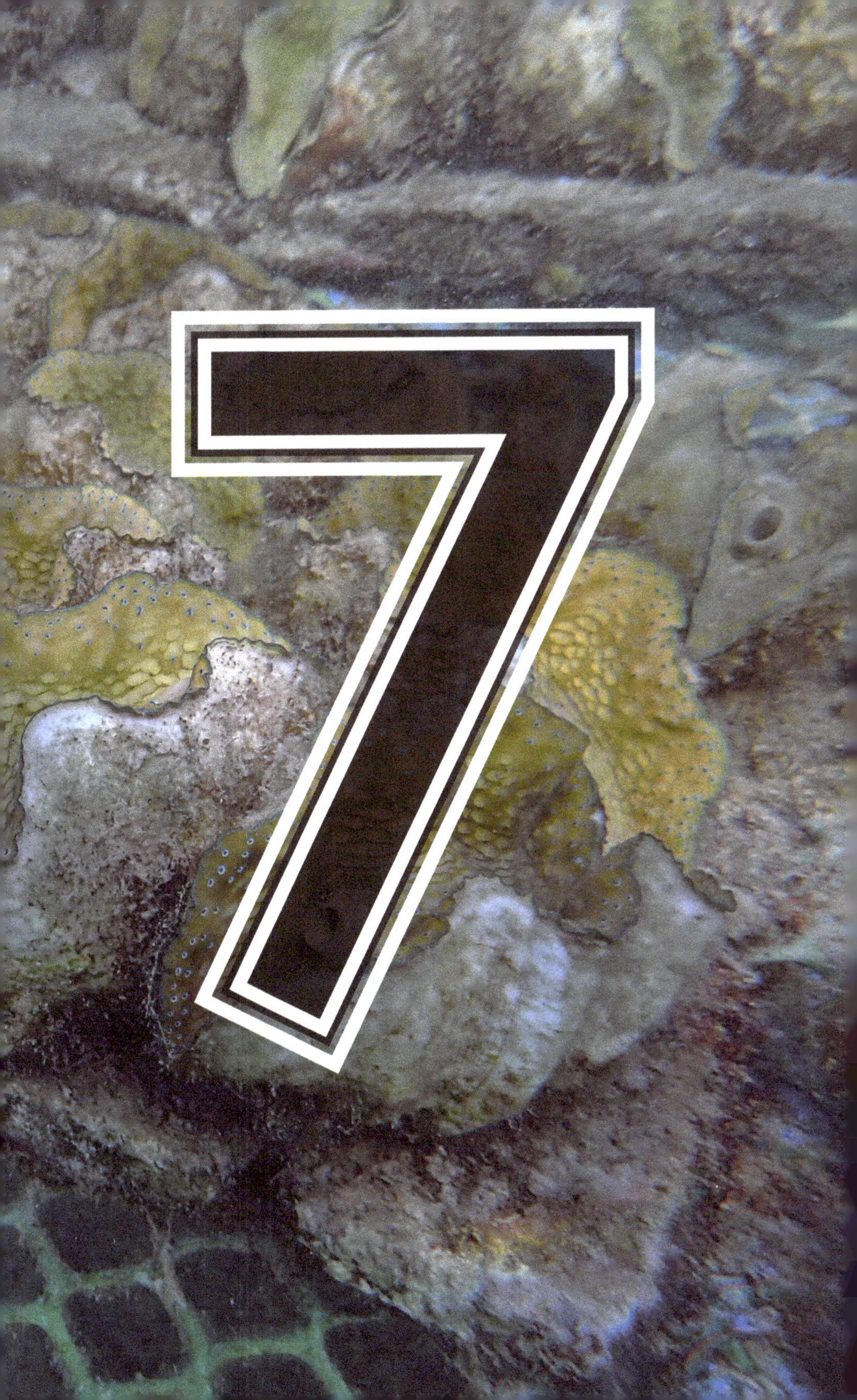
7

7. Conservation Approaches

Around the late 1980s, conservationists developed concerns over the status of giant clams in the wild after reports confirmed that the intense and large-scale exploitation had severely decimated wild populations across the Indo-Pacific. As a result, their dire situation prompted the relevant stakeholders, such as local communities, fishers, traditional owners, non-governmental organisations and academics, and the government authorities, to act and prevent the further collapses of giant clam populations. Broadly, various measures targeted at the global, regional and local giant clam populations have helped raise awareness of the threats they face, better regulate trade and mitigate the decline of existing populations. In addition, decades of giant clam research have also played a part in our understanding of their systematics, biology, physiology and ecological significance, reinforcing the importance of protecting these charismatic molluscs. Henceforth, we will further examine some key conservation concerns and approaches directed at giant clam populations globally, regionally and locally.

Regulating International Wildlife Trade

The regulation of international wildlife trade falls under the purview of CITES, which stands for the Convention on International Trade in Endangered Species of Wild Fauna and Flora. CITES is a global Treaty among governments to ensure that the international trade in specimens of wild animals and plants does not threaten the survival of these species. CITES does so by listing species in one of the three Appendices according to the

degree of protection and controls they need and administering and authorising the trade of species covered by the Convention through a licensing system. The Convention officially has 184 Parties, which comprises 183 States and one regional economic integration organisation, the European Union, for which the Convention has entered into force. The Parties to CITES are collectively referred to as the Conference of the Parties (COP), where the latter meets once every two to three years to review the implementation of the Convention, such as evaluating the progress of species conservation included in the Appendices and considering proposals to amend the species lists in Appendices I and II.

Since the early 1980s, the giant clams have been protected under CITES Appendix II, which consists of species that are not necessarily currently threatened with extinction but may become so unless trade is closely controlled. These charismatic bivalves were first highlighted as species to be included in CITES due to the rapid overexploitation of

their wild populations. *Tridacna gigas* and *Tridacna derasa* were listed in 1983, while the remaining members of the subfamily were listed in 1985 based on the 'look-alike' principle (i.e., taxa deemed to look similar are listed together within the same Appendix if one or more of the taxa are threatened through international trade). All living or dead specimens of giant clams, including all readily identifiable parts and derivatives, are subject to the Treaty's provisions for Appendix II species. CITES also clearly states that international trade in giant clams is permitted only if the respective CITES Parties issue relevant export and import certifications. In some countries, CITES has also been referred to as an authoritative source for identifying endangered giant clam species that subsequently informs their local laws focused on biodiversity protection.

The main strengths of the CITES trade data and reporting process for the giant clams include a high level of awareness that CITES trade controls are in place for these bivalves since their listing in 1985, as well as a good level of documentation concerning the trade in giant clam commodities from certain specific countries such as Viet Nam and France. However, a major shortcoming is that CITES enforcement largely depends on whether the countries involved in the trade are Parties to the Treaty or if a non-Party is trading with the Parties. Historically, several countries, such as Taiwan and Maldives, formerly heavily involved in trading giant clams, were not CITES Parties, so their trade may not have been reported to assess trade levels accurately. Another potential issue with CITES is that the scope of the Treaty does not apply to the domestic trade of giant clams within a country, regardless of its status as a Party member. This means the collection and trade of giant clam products may be allowed, depending on the country's local laws and regulations.

On the other hand, CITES conducts regular workshops and other initiatives to strengthen capacity and implementation for both the Parties and non-Parties to the Convention. For example, beyond the COP meetings, the CITES Animals Committee holds regular meetings to review the trade levels of listed species. In addition, it provides timely updates to Parties, such as the species' taxonomy and nomenclature. These events ensure that Parties use accurate information to monitor their trade levels without compromising the survival of Appendix-listed species.

IUCN Red List of Threatened Species

The International Union for Conservation of Nature (or IUCN) is an international organisation seeking to promote nature conservation and the sustainable use of natural resources. The IUCN Red List of Threatened Species was established in 1964 as a one-stop information source on the global extinction risk status of flora and fauna to encourage well-informed conservation decisions and policy changes. More than 150,000 species have been assessed for the IUCN Red List, with continual assessments of newly validated species and the re-assessments of status for some existing species. Specifically, the information on the species' geographic range, population size, habitat and ecology, use and/or trade, threats and conservation actions are assessed, peer-reviewed and published. Recent assessments of species status are based on the 2001 IUCN Red List Categories and Criteria: Version 3.1, where the species are broadly placed in one of nine categories: Not Evaluated, Data Deficient, Least Concern, Near Threatened, Vulnerable, Endangered, Critically Endangered, Extinct in the Wild and Extinct. Information on the IUCN Red List can direct us on where and what actions are needed to protect critical species from extinction and provide a means to factor biodiversity needs into decision-making processes. The IUCN Red List is widely used by government agencies, wildlife conservation practitioners, educational organisations and the business community so that it can influence various conservation interests. This may include guiding scientific research, informing conservation planning and policy, influencing resource allocation, and raising environmental awareness and education literacy.

The first IUCN Red List assessment for seven species of giant clams was published in the 1983 IUCN Invertebrate Red Data Book. To assess their risks of extinction, the book contained information on the giant clam species' range, ecology, scientific interest, threats and conservation measures. Only *Tridacna gigas* and *Tridacna derasa* were listed as Vulnerable in the assessment. In contrast, the remaining species (*Hippopus hippopus*, *Hippopus porcellanus*, *Tridacna crocea*, *Tridacna maxima* and *Tridacna squamosa*) were listed as either Indeterminate or Insufficiently Known. Subsequently, these bivalves were regularly re-assessed between 1986 and 1996 by experts from the IUCN Species Survival Commission (SSC) Mollusc Specialist Group. Their statuses had remained unchanged until the last assessment in 1996. This latest assessment had also included two other species: *Tridacna rosewateri* and *Tridacna mbalavuana*. As this evaluation was carried out before 2001, the assessors had used the 1994 IUCN Red List Categories and Criteria: Version 2.3, where the Red List category for Lower Risk was still applied. In the 1996 assessment, four species were listed as Vulnerable (VU): *Tridacna gigas*, *Tridacna derasa* and *Tridacna rosewateri*, based on the rate of decline of remaining wild stocks, and *Tridacna mbalavuana*, on the basis of its small and declining area of suitable habitats. *Hippopus hippopus*, *Hippopus porcellanus*, *Tridacna maxima* and *Tridacna squamosa* were listed as Lower Risk/conservation dependent (LC/cd) on the basis of the decline and disappearance of many populations. *Tridacna crocea* was listed as Lower Risk/least concern (LR/lc) as the taxa was considered of lesser concern compared to the rest of giant clam species in the other threat categories, but this does not mean that it was of no conservation concern.

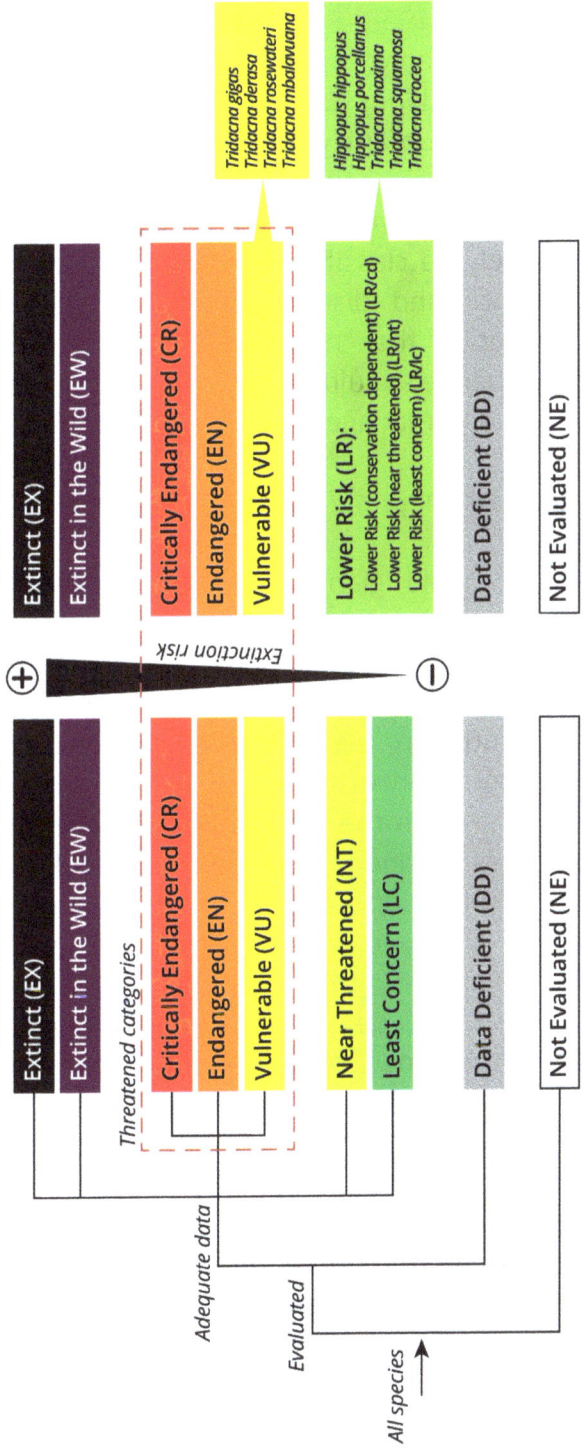

Version 3.1 (IUCN 2001)

- Extinct (EX)
- Extinct in the Wild (EW)

Threatened categories

- Critically Endangered (CR)
- Endangered (EN)
- Vulnerable (VU)
- Near Threatened (NT)
- Least Concern (LC)
- Data Deficient (DD)
- Not Evaluated (NE)

Adequate data
Evaluated
All species

Version 2.3 (IUCN 1994)

- Extinct (EX)
- Extinct in the Wild (EW)

Extinction risk (+ / –)

- Critically Endangered (CR)
- Endangered (EN)
- Vulnerable (VU)
- Lower Risk (LR):
  Lower Risk (conservation dependent) (LR/cd)
  Lower Risk (near threatened) (LR/nt)
  Lower Risk (least concern) (LR/lc)
- Data Deficient (DD)
- Not Evaluated (NE)

Tridacna gigas
Tridacna derasa
Tridacna rosewateri
Tridacna mbalavuana

Hippopus hippopus
Hippopus porcellanus
Tridacna maxima
Tridacna squamosa
Tridacna crocea

The IUCN Council first adopted the IUCN Red List categories (Version 2.3) in December 1994, which became widely recognised internationally. Version 2.3 was used for the 1996 IUCN Red List of Threatened Animals, which included the giant clams. Following feedback from IUCN and SSC members, the IUCN Council revised and adopted the latest version of the IUCN Red List Categories (Version 3.1) in February 2000. All new assessments from January 2001 now use the latest adopted version. Therefore, the current Red Listings for giant clams are based on Version 2.3, and a re-assessment is urgently needed using Version 3.1.

Even though the 1996 Red Listings for the giant clams had been considerably comprehensive back then, it now urgently requires updating to reflect the contemporary status quo of taxa in the subfamily. For instance, the recently rediscovered species, *Tridacna squamosina*, *Tridacna noae* and *Tridacna elongatissima*, have yet to be assessed for the Red List. Still, they will likely be classified as Data Deficient since appropriate abundance and distributional data still need to be included. On top of that, new threats and challenges arising from climate change should be considered to guide the relevant conservation and management of giant clams.

Local Management and Enforcement Efforts

In areas where wild giant clams can be found in the Indo-Pacific region, the primary focus of giant clam conservation and management at the local or regional scales is lowering exploitation rates. Relevant legal measures were examined and presented in the following overview tables to understand the protection extended to these giant clams in the respective countries in the Indo-Pacific. In general, the types of legal measures the Indo-Pacific countries took would depend on the local utilisation patterns of giant clams as a coastal resource and the type and extent of threats these shellfish face. Most of the countries in the regions of Southeast Africa, East Asia and Southeast Asia examined have listed their giant clams and their species as protected species under the relevant laws related to fisheries or wildlife conservation. These protection laws may be accompanied by other specific legal measures to guide the harvest, sale or export of wild giant clams in the respective countries. In some of the countries examined (i.e., Djibouti, Egypt, Eritrea, Jordan, Sudan, Yemen, Republic of Kiribati, and the Northern Mariana Islands), there are also non-specific mentions of giant clams being protected by their law as they are often defined together with other marine aquatic species in categories such as fish, shellfish, seashells or shells.

On the other hand, the Nairobi Convention for the Protection, Management and Development of the Marine and Coastal Environment of the Eastern African Region is an example of seeking regional cooperation from the Contracting Parties (Comoros, France, Kenya, Madagascar, Mauritius, Seychelles, Somalia, Tanzania and the Republic of South Africa) to address the protection of marine resources. The Convention was signed in 1985 and came into force in 1996, with amendment and adoption in April 2010. Under this Convention, the *Tridacna maxima* giant clam species is listed as a "Species of Wild Fauna Requiring Special Protection", but no exact measures have been proposed on how to protect and manage this species.

As the purpose of harvesting giant clams is wide-ranging such as for subsistence, commercial, or other individual uses, including recreational, tourism and aquaculture, the legal measures vary considerably across the Indo-Pacific countries. Furthermore, the degree of managing giant clam harvesting could range from introducing restrictions to a full-out ban. For instance, in the Pacific Islands, where giant clams are a crucial coastal food source for the local communities, they generally restrict harvesting their wild giant clams to alleviate fishing pressures. These restrictions may include setting minimum size limits for harvesting, imposing harvesting quotas or bag limits, restricting clam fishing to free diving only, and banning mechanical fishing equipment. In addition, the giant clam aquaculture industry has a strong presence in the Pacific Islands, and there are exceptions stated in the laws to allow the sale and export of only cultured specimens. Interestingly, none of the Pacific Islands is found to introduce bans on harvesting wild giant clams for subsistence use, but some countries have banned the commercial harvest, sale or

An Overview of the Relevant Legal Measures Relating to the Conservation and Management of Giant Clams Across the Indo-Pacific

Region	Red Sea		South-east Africa										Indian Ocean		East Asia		
Country / Legal Measures	Saudi Arabia	Somalia	Comoros	Kenya	Madagascar	Mauritius	Mayotte	Mozambique	La Réunion	Seychelles	South Africa	Tanzania	India	Maldives	China	Japan	Taiwan
Listing giant clams and their species as protected species in legislation		√	√	√	√	√	√	√	√	√	√	√	√		√		√*
Considered an offence to fish, buy, sell, import or export wild giant clams and is punishable by law															√		√*
Introduced restrictions on harvesting wild giant clams (such as using size, weight or bag limits, gear restrictions or permits)																√†	
Introduced bans on harvesting wild giant clams	√									√					√‡		√*
Introduced restrictions to manage commercial harvest, sale or export of wild giant clams (such as using quota, exclusions or permits)										√					√		
Banning the commercial harvest, sale or export of wild giant clams and their derivatives													√	√	√‡		√
Introduced restrictions on individual uses, including recreational, tourism and aquaculture																√†	

Region	Southeast Asia								Melanesia					
Country / Legal Measures	Brunei Darussalam	Indonesia	Malaysia	Myanmar	Philippines	Singapore	Thailand	Viet Nam	Australia	Fiji	New Caledonia	Papua New Guinea	Solomon Islands	Vanuatu
Listing giant clams and their species as protected species in legislation		✓	✓‡	✓	✓	✓	✓	✓			✓		✓	
Considered an offence to fish, buy, sell, import or export wild giant clams and is punishable by law		✓	✓		✓	✓	✓							
Introduced restrictions on harvesting wild giant clams (such as using size, weight or bag limits, gear restrictions or permits)	✓	✓							✓	✓	✓	✓		
Introduced bans on harvesting wild giant clams				✓	✓		✓							
Introduced restrictions to manage commercial harvest, sale or export of wild giant clams (such as using quota, exclusions or permits)						✓			✓	✓				
Banning the commercial harvest, sale or export of wild giant clams and their derivatives		✓		✓	✓						✓	✓	✓	✓
Introduced restrictions on individual uses, including recreational, tourism and aquaculture		✓							✓	✓	✓		✓	✓

^Laws apply only in Hainan Province.
†Laws apply only to Okinawa Prefecture.
*Laws apply only to Taitung and Penghu Counties.
‡Law applies only to the established marine park or marine reserve in Malaysia.

Region	Micronesia					Polynesia								
Country	Federated States of Micronesia	Guam	Marshall Islands	Nauru	Palau	American Samoa	Cook Islands	French Polynesia	Pitcairn Islands	Niue	Samoa	Tokelau	Tonga	Tuvalu
Legal Measures														
Listing giant clams and their species as protected species in legislations									✓					
Considered an offence to fish, buy, sell, import or export wild giant clams and is punishable by law	✓								✓					
Introduced restrictions on harvesting wild giant clams (such as using size, weight or bag limits, gear restrictions or permits)	✓	✓			✓	✓	✓	✓		✓	✓	✓	✓	
Introduced bans on harvesting wild giant clams														
Introduced restrictions to manage commercial harvest, sale or export of wild giant clams (such as using quota, exclusions or permits)	✓	✓			✓	✓		✓						✓
Banning the commercial harvest, sale or export of wild giant clams and their derivatives			✓				✓						✓	
Introduced restrictions on individual uses, including recreational, tourism and aquaculture		✓	✓	✓	✓			✓					✓	✓

export of wild giant clams. In contrast, more countries in East Asia and Southeast Asia prohibit the collection of wild giant clams for subsistence and commercial uses, which may be punishable by law in some of these countries.

By and large, it is encouraging that most Indo-Pacific countries recognise their giant clams as a group of species requiring strategic protection and management. On the other hand, these legal measures are only effective if they are appropriately implemented with active interventions to reduce their overexploitation and prevent the collapse of wild populations. Achieving effective law enforcement would depend on several factors, such as the present levels of exploitation in the country (for instance, a highly exploited population will require more intensive policing), the capacity to enforce these laws, and the community's cooperation and cooperation willingness to adhere to regulations. However, other challenges at large can impede law enforcement, such as the difficulty of enforcing specific unfeasible regulations and the widespread problem of poaching. For example, enforcing the rule on clam size limits in Tonga has proven difficult without the full cooperation of fisherfolk and retailers. In this case, it took a lot of work to cover all the markets comprehensively to check clam sizes, the inaccuracy of measuring the clams sold frozen in plastic bags and the loose rule where fisherfolk can consume undersized clams at home while selling only the legal-sized ones. Also, the increased prevalence of illegal clam fishing in the region complicates the already arduous task of managing giant clam populations in the wild. While the law enforcement of legal measures is a challenging aspect of species conservation, the continued implementation of these local-scale management strategies to manage and protect wild giant clam populations is crucial as they support the global call for their conservation.

Mariculture and Restocking Interventions

The rapid decline of natural giant clam populations due to intensive overexploitation in the 1960s spurred the advancement of mariculture techniques to breed giant clams as a way to increase individual numbers quickly. Unlike other commercially important marine molluscs such as oysters and mussels, the breeding of giant clams is a relatively recent development that has experienced rapid progress since its initiation in the 1970s. Mariculture of giant clams was pioneered in the 1970s at the University of Guam Marine Laboratory in Guam and the Micronesian Mariculture Demonstration Centre (MMDC) in Palau. This breeding effort was further supported by the Australian Centre for International Agricultural Research (ACIAR) in Australia in the 1980s and consolidated by ICLARM (now known as WorldFish) in the Solomon Islands in the late 1980s and early 1990s. These programmes subsequently supported extensive mariculture research and technical training

throughout the Pacific and Southeast Asia. The growing confidence in the mariculture capabilities of giant clams presented opportunities in the Indo-Pacific region to mass produce cultured individuals for purposes in conservation, research, domestic consumption, and commercial use, such as the marine ornamental trade. In particular, mariculture offers the chance to use cultured juvenile giant clams for restocking rare species or extirpated populations. Although the mariculture of giant clams does not have apparent or major harmful effects on the environment, the accidental introduction of exotic parasites, diseases, and other 'hitch-hiking' biota remains a possibility; this may be overcome if the imported giant clams undergo appropriate quarantines. Therefore, mariculture for restocking giant clams can be an immediate remedy for declining wild populations since it would take decades for natural giant clams to recover and return to their original states.

According to the book 'Saving Giants' by Gerald Heslinga, one of the pioneers in developing the giant clam mariculture capacity in the 1970s, he wrote that there were at least 34 functioning giant clam hatcheries across 25 countries (accurate as of 2016). Many of these giant clam hatcheries typically operate on some commercial (or semi-commercial) basis. They may also function as a means to support conservation and facilitate sustainable harvesting through foreign aid support or other subsidies. Among some community-led giant clam grow-out operations, these mariculture activities can provide sustainable livelihood opportunities as long as projects are run as sustainable and cost-effective enterprises, benefiting communities living in remote areas such as atolls and outlying islands where job options are limited. Mariculture for restocking giant clams has been widely adopted around the Indo-Pacific region. In the 1980s and 1990s, many cultured juvenile giant clams were shipped to the region, matured in ocean nurseries at the destination countries, and later used as breeding stocks in local hatcheries. Sadly, the outcomes of most restocking initiatives have neither been well-studied nor well-documented. These restocking programmes often do not have protocols for fisheries officers and managers to follow, as well as lacking regular monitoring to ascertain the success of such efforts over time. The survivorship of restocked clams also varies widely within and among countries, with the leading causes of mortality being predators, storms, poaching, and the lack of funding for continuous husbandry. A study found that hatchery-bred clams, spawned from an inherently sub-sampled population (broodstock), may be less genetically diverse, which could increase their vulnerability to diseases and environmental stressors. Other challenges that clam breeders face are operations-related, such as high mortality rates, high running costs and labour-intensive rearing of juvenile clams to reach escape size.

Even though the operators of giant clam mariculture face obstacles such as the upkeep of intensive farming and waning funds, it remains the most feasible and prompt solution to slowing down declines in the wild. Beyond restocking to boost giant clam numbers, the other desired outcome is the self-recruitment by these restocked clams. More than 20 years after restocking, there is now firm evidence in Yap (Federated States of Micronesia) and the Philippines, where the restocked *Tridacna derasa* and *Tridacna gigas*, respectively, produced recruits within the vicinity of restocking sites. Similarly, the local community of Shiraho Village on Ishigaki Island of Okinawa Prefecture had carried out a stock enhancement effort (locally known as *Satoumi*), where 9,100 cultured juvenile *Tridacna crocea* were purchased from the Okinawa government and restocked in 2009 and 2010 under the guidance of fisheries extension officers. These restocked clams were introduced as spawning stocks to help increase giant clam numbers in the surrounding area and provide new tourist attractions. Notably, these places are also where the mariculture and restocking efforts have maintained momentum for more than 30 years, suggesting that programme longevity and commitment are key factors contributing to the success of restocking initiatives. These encouraging stories confirm that restocking programmes can achieve reproduction sustainability and that creating new generations can benefit local communities. Overall, the restocking of cultured giant clams is not without its challenges, but novel mariculture techniques are emerging rapidly to improve the robustness of bred clams. These considerations include the selective cross-breeding of tolerant holobionts and inoculating tolerant strains of endosymbiotic microalgae to larvae. Furthermore, the five decades of experience in this conservation endeavour have shown that the most reliable strategy is implementing well-managed, financially sustainable giant clam hatcheries and well-defended ocean nurseries.

The Bolinao Marine Laboratory of the University of the Philippines has been culturing giant clams to replenish locally since 1985. Over 20,000 *Tridacna gigas* have been restocked at more than 40 sites nationwide. The recent discovery of juvenile *Tridacna gigas* suggests that restocked individuals can likely restock local populations, and this is a remarkable achievement for the Philippines' giant clam conservation effort. ©Mei Lin Neo

Giant clams are relatively long-lived animals on coral reefs. Therefore, the long-term monitoring of their presence and health condition on the reefs is a reflection on the state of the environment, thus making them valuable and vital indicators of coral reef health.

Conservation Prioritisation using Biodiversity Measures

Marine biodiversity conservation prioritises preventing the loss of species that maintains community assembly and ecosystem functioning. Most of the current conservation efforts and strategies in the marine environment generally focus on protecting areas with high numbers of species (or flagship species) as a way to preserve ecosystem integrity and increase the success of conservation initiatives. But to rely on species richness and patterns as a gauge for prioritising conservation is inadequate, as the use of taxonomic diversity alone cannot account for the individual species' variability in terms of their uneven functional contributions and evolutionary histories within a community. As a result, these approaches may unintentionally underrepresent other components of biodiversity, such as functional and phylogenetic diversity. Hence, there has been a shift in perspectives on how prospective conservation prioritisation strategies should use multiple biodiversity measures to assess the relative contributions of species assemblages towards the functioning of ecosystems. In particular, maximising phylogenetic diversity has been widely advocated as a conservation strategy to preserve evolutionary history for future generations. It currently forms the basis for global conservation frameworks such as the Evolutionarily Distinct and Globally Endangered (EDGE) of Existence programme. The use of functional traits as a metric to inform conservation has also been advocated by the IUCN Red List of Ecosystems and tested in studies to ensure the resilience and functions of ecosystems are considered and maintained.

Among the marine taxa, giant clams have received considerable attention calling for their increased protection over the past decades. So far, most of the existing conservation efforts for giant clams have been centred on protecting individual species based on the extent of localised threats in the respective countries or regions. However, these conservation approaches discussed earlier in this Chapter have yet to directly consider or incorporate other facets of biodiversity or the ecological values of species with respect to ecosystem functioning for conservation purposes. Also, the resources and funds supporting the conservation of giant clams in recent years have been waning. Furthermore, they are usually short-term (i.e., less than five years) despite the global threats and challenges. There is an urgent need to trial novel conservation prioritisation methods to raise the profiles of giant clams in regional and international marine conservation discussions. A published study led by Edwin Tan and colleagues set out to test this conservation prioritisation approach based on assessing various biodiversity components associated with taxonomic, functional and evolutionary measures of giant clam assemblages. Specifically, they wanted to determine the efficacy of three biodiversity measures in identifying potentially threatened species from a molluscan subfamily of 12 species that contribute disproportionately to functional and phylogenetic diversity and highlight areas for global conservation prioritisation.

The underpinning information used for the analyses was the biological and ecological data that characterise the individual giant clam species, coupled with the well-supported phylogenetic data discussed in Chapter 2. The study first quantified the taxonomic, functional and phylogenetic diversity measures for 12 giant clam species throughout 25 marine provinces (see the map on taxonomic diversity in Chapter 1). Then it evaluated the strength of relationships among diversity measures by comparing whether the biodiversity patterns corresponded. The specific biodiversity measures used were species richness (SR), functional richness (FRic) and Faith's phylogenetic diversity (PD_F), where they are analogous to taxonomic, functional (FD) and phylogenetic diversity (PD). The results revealed positive correlations among these three biodiversity measures, with the most robust relationship between FRic and PD_F. The spatial patterns of FD and PD for the giant clams are most congruent but less so with SR patterns. This suggests that conserving giant clam species based on SR alone can only partially incorporate the contributions derived from FD and PD.

The spatial patterns of ND_{FD} and NF_{PD} further revealed that the provinces (i.e., Bay of Bengal, Andaman and numerous parts of the eastern Indo-Pacific) bordering some of the highest SR areas within the Central Indo-Pacific are found to support more functionally and evolutionary distinct giant clam assemblages than expected for their lower richness. In particular, the presence of certain species in provinces, such as the *Tridacna gigas* and *Hippopus* spp. could boost FD and PD levels, respectively, as these species have the most significant functional and evolutionary diversity contributions within community assemblies (defined here as marine provinces), respectively. Based on this study, we can further highlight overlooked areas that warrant more attention by considering the contributions of giant clam assemblages across multiple biodiversity components. This study's results also provide comprehensive biodiversity information regarding the giant clam species for managers and decision-makers to initiate measures to holistically capture species' ecological and evolutionary standings across global reef ecosystems.

While these findings underscore the importance of integrating functional trait contributions and evolutionary diversity into conservation planning to better differentiate giant clam assemblages in the Indo-Pacific region, the primary focus of many existing marine conservation efforts is protecting reef-building coral-rich areas such as the Coral Triangle. Given that the giant clams and corals are vastly different in their contributions towards the ecosystem functioning and evolutionary history, their spatial patterns of biodiversity are highly contrasting, which makes it challenging to conserve both taxa equally. Even though it is evident that protecting reef-building corals would be insufficient to capture the overall functional and evolutionary contributions of the numerous groups of reef taxa that inhabit the same ecosystem, it remains most practical to prioritise efforts in coral-rich regions considering the limited resources available and the diverse competing priorities ranging from socio-economic to ecological needs. Even so, there are areas within the Coral Triangle, such as the Andaman and Tropical Southwestern Pacific, that hold disproportionately high ND_{FD} and ND_{PD} in giant clams that could be afforded increased protection.

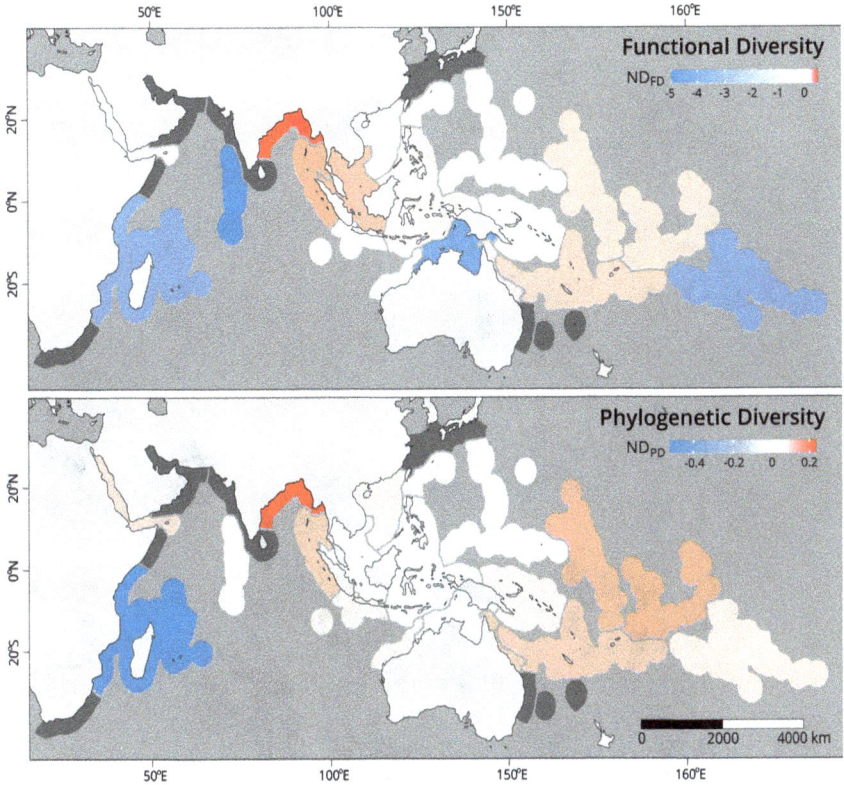

Spatial patterns of functional and phylogenetic diversity of giant clam species in the Indo-Pacific. The normalised difference (ND) was obtained for each province by quantitatively comparing differences between observed and null values for each biodiversity component based on whether each province held more (or less) functional and phylogenetic diversity than random expectation. A positive ND would suggest a more diverse assemblage of giant clam species than expected. In contrast, a negative ND value would indicate that the diversity of species was lower than expected. Overall, there is strong congruence between spatial patterns of ND_{FD} and ND_{PD} for giant clam assemblages, but high ND_{FD} and ND_{PD} provinces do not necessarily hold high species richness. Dark grey areas indicate provinces with fewer than three species were omitted from the analyses. Reproduced from Tan et al. (2022).

8. How To Identify Species

In this book, we compile and feature the most updated information on the currently recognised giant clam species. It contains a comparative table with anatomical characters of the shell, outer mantle, and ecology to provide an overview of general differences across the species. This is followed by the individual species pages, which include detailed information on the species biology, geographic distribution, taxonomy and key identifying features that can be used for identification. Finally, a comprehensive dichotomous key is also included in this book to assist you in a step-by-step approach to identifying and differentiating the giant clam species.

In most cases, the shell morphology is often diagnostic for the identification and differentiation between giant clam species. However, the shells may be less readily observable in the field as the external surfaces of shell valves are often encrusted with marine organisms that obscure the shell features. In addition, while the highly visible outer mantle has several discernible features that can be useful in narrowing down potential species, do note that the colouration and patterns of mantles are not useful traits because they are highly variable and plastic within and between giant clam species. Therefore, identifying giant clam species, especially those highly similar-looking species, requires using multiple characteristics from the shell and outer mantle and other information, such as geographic distribution, to determine species identity. All in all, the information provided in this book will allow you to readily recognise the giant clam species by examining photographs or observing the animals *in situ*.

How to Use the Species Page

A. *Species name*
The recognised species name.

B. *Author citation*
This refers to the person who first described the species and the year it was described.

C. *General species information*
This section elaborates on its broad geographic occurrence and maximum body size in cm.

D. *Common name(s)*
A name or names of an organism that is known to the general public or is based on the normal language of everyday life.

E. *Biology*
This section elaborates on the species' habitat, ecology, and behaviour in its natural environment, where information is available.

F. *Geographic distribution*
This refers to the natural occurrence of the species. Do note that this distribution does not imply a species can be found in all locations within shading but that if conditions are suitable, it might occur there.

G. *Taxonomy and morphology*
This section primarily elaborates on the key characteristics that can help to distinguish from all other species. It may also contain distinctive features to help differentiate from other similar-looking species.

H. *Illustrations of shell valves, scale (cm)*
A detailed hand illustration of species shell valves showing the dorsal and lateral views.

I. *Depth occurrence (m)*
The maximum depth in m that the species is known to occur.

A. *Species name*
B. Author citation

C. General species
information

D. Common name(s)

E. Biology

F. Geographic distribution

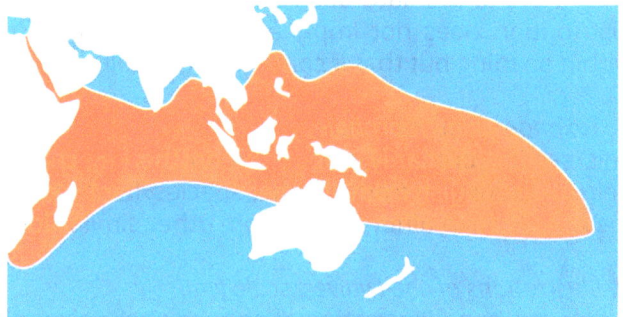

G. Taxonomy and
morphology

H. Illustrations of shell
valves, scale (cm)

I. Depth occurrence (m)

Comparative Table of Species

Genus	Hippopus		Tridacna		
Species	*hippopus*	*porcellanus*	*gigas*	*mbalavuana*	*derasa*
Maximum shell length (cm)	50.0	41.1	137.0	56.0	60.0
Shell symmetry	Inequilateral	Inequilateral	Equilateral	Inequilateral	Inequilateral
Shell rib depth	Deep	Shallow	Deep	Shallow	Shallow
Scutes in adults	Absent	Absent	Absent	Absent	Absent
Hinge length : Shell length	>0.5	>0.5	>0.5	>0.5	>0.5
Umbo position	Posterior	Posterior	Central	Posterior	Posterior
Byssal orifice size	Narrow	Narrow	Small	Narrow	Small
Mantle extension beyond shell margins	No	No	Yes	No	Yes
Incurrent siphon with guard tentacles	Absent	Present	Absent	Present	Present
Hyaline organs	Absent	Absent	Present	Absent	Present
Attachment to reef	Free-living	Free-living	Free-living	Free-living	Free-living
Maximum depth occurrence (m)	10.0	9.0	10.0	33.0	20.0

Tridacna

crocea	squamosa	noae	maxima	rosewateri	squamosina	elongatissima
15.0	47.6	28.0	50.0	26.0	32.0	33.5
Inequilateral	Equilateral	Inequilateral	Inequilateral	Inequilateral	Inequilateral	Inequilateral
Shallow	Deep	Moderate	Moderate	Deep	Deep	Moderate
Present	Present	Present	Present	Present	Present	Present
<0.5	=0.5	<0.5	<0.5	<0.5	<0.5	<0.5
Anterior	Central	Anterior	Anterior	Posterior	Central	Anterior
Wide	Small	Wide	Wide	Wide	Wide	Wide
Yes	Yes	Yes	Yes	Yes	Yes	Yes
Present	Present	Present	Present	Present	Present	Present
Present	Present	Present	Present	Present	Present	Present
Embedded	Free-living	Embedded	Embedded	Free-living	Free-living	Embedded
10.0	42.0	21.9	21.2	13.0	5.0	15.0

©Li Keat Lee

Hippopus hippopus
(Linnaeus, 1758)

Hippopus hippopus is found throughout the Indo-Pacific, except for the Red Sea and Western Indian Ocean. It can grow to 40 cm long, with the largest specimen recorded at 50 cm.

Common names

Horse's Hoof Clam, Strawberry Clam

Biology

This species often inhabits the shallow, nearshore patches of reef, sandy areas and seagrass beds that can be exposed during low tides. It is occasionally found as deep as 10 m. Byssal attachment is present in juveniles, but older ones mainly lie free-living (unattached) on the substratum.

Geographic distribution

Taxonomy and morphology

The shell is thick and heavy, has strong radial ribbing with reddish blotches in irregular bands, and has a narrow byssal orifice with interlocking teeth. Mantle usually exhibits green, yellow-brown or grey-mottled patterns and does not extend beyond the upper shell margins. Hyaline organs are lacking, and the incurrent siphon bears no guard tentacles.

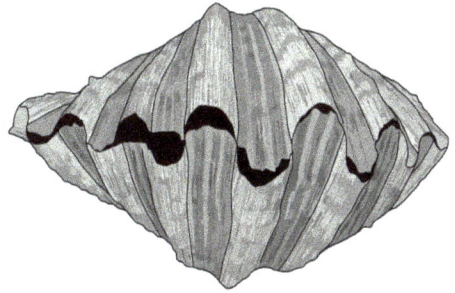

Depth occurrence

10 m

New Caledonia, Palfrey

Malaysia, Sibu-Tinggi

Papua New Guinea

Malaysia, Sibu-Tinggi

Indonesia, Papua

Indonesia, Kei Island

Philippines, Bolinao

Palau, Kayangel

115

Hippopus porcellanus
Rosewater, 1982

Hippopus porcellanus is specifically known only from the Sulu Archipelago and Palawan (Philippines), Sabah (Malaysia), Sulawesi and Raja Ampat (Indonesia), Palau, and Milne Bay Province (Papua New Guinea). It typically reaches approximately 40 cm long, with the largest specimen at 41.1 cm.

Common name

China Clam

Biology

This species is usually found free-living on intertidal reef flats and the shallow reefs along the edges of lagoons.

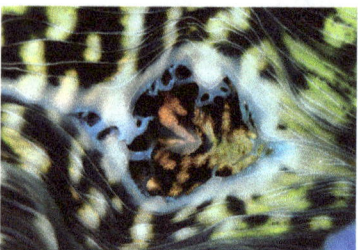

Geographic distribution

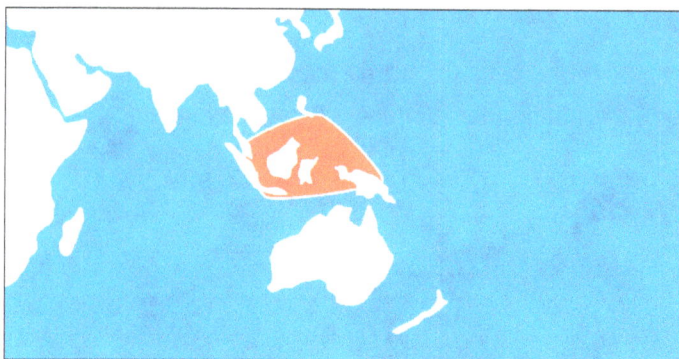

Taxonomy and morphology

The shell is smoother and thinner, has a more rounded oval shape than *Hippopus hippopus*, and has a narrow byssal orifice with interlocking teeth. Mantle is generally grey or brown and does not extend beyond the upper shell margins. Hyaline organs are lacking, and the incurrent siphon bears prominent indistinct guard tentacles. This species may sometimes be mistaken for *Tridacna derasa* due to their similar shell shape and texture.

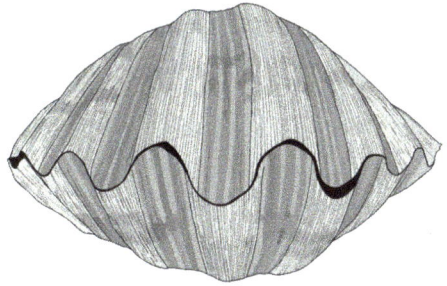

Depth occurrence

10 m

East Malaysia, Sabah

118

Philippines, Palawan

Philippines, Palawan

119

Tridacna gigas
(Linnaeus, 1758)

Tridacna gigas is found throughout the Indo-Pacific, from Myanmar to the Republic of Kiribati (but not the Cook Islands) and the Ryukyus (southern Japan) to Queensland (Australia). It is the only truly gigantic giant clam species, with the largest specimen recorded at 137 cm long and the heaviest specimen recorded at 500 kg.

Common name
True Giant Clam

Biology
This species is usually free-living on either sand or hard reef substrata, and typically lives in coral reefs with good light penetration.

Geographic distribution

Taxonomy and morphology

The shell is elongate-oval to fan-shaped, very thick and heavy with a small byssal orifice. It is easily recognised by its large body size and distinctive elongate, triangular projections on the upper shell margins. Mantle exhibits mostly dull brown and olive-green colours, and the mantle edge bears numerous hyaline organs bordered by iridescent blue-green circles. Unlike the other *Tridacna* species, the incurrent siphon of *Tridacna gigas* bears no guard tentacles.

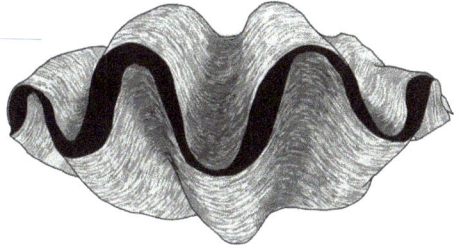

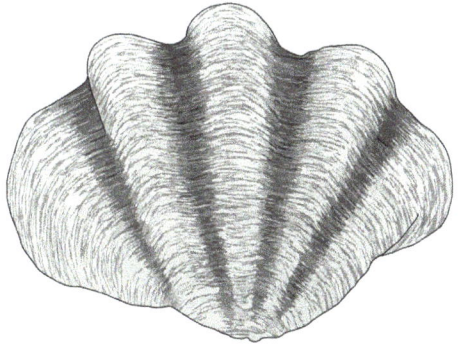

Depth occurrence

10 m

Indonesia, Raja Ampat, Kri Island

Philippines, Palawan

East Malaysia, Sabah

124

New Caledonia, Mermaid Cove

Indonesia, Bunaken Island

Indonesia, Papua

Myanmar, Rakhine, Andrew Bay

Palau, Ebiil

Papua New Guinea

Close-up of the hyaline organs

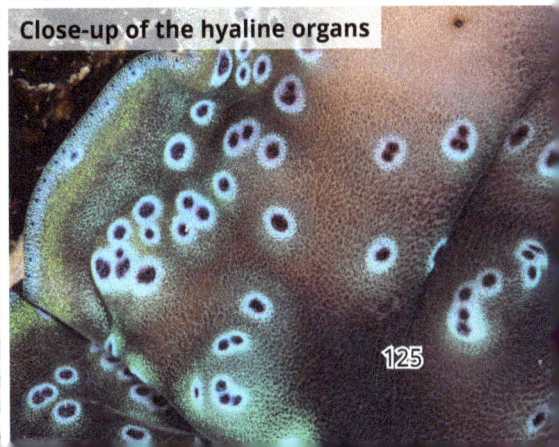

125

Tridacna mbalavuana
Ladd, 1934

Tridacna mbalavuana is known only from Fiji, Tonga, New Caledonia, and Australia. It is generally rare throughout its known range. It typically reaches approximately 50 cm long, with the largest specimen recorded at 56 cm. Previously recognised as "*Tridacna tevoroa*".

Common name

Devil Clam

Biology

This species usually found in relatively deep waters (>20 m) compared to other giant clam species, and may be intolerant of conditions in shallow water.

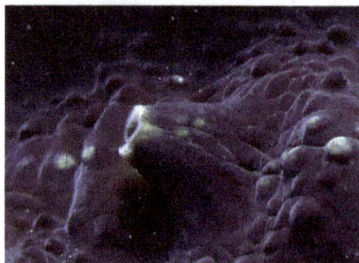

Geographic distribution

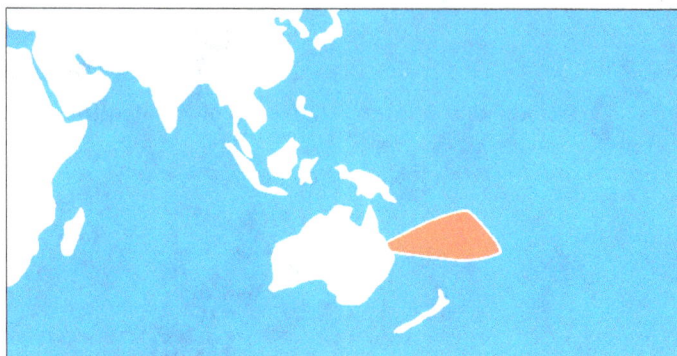

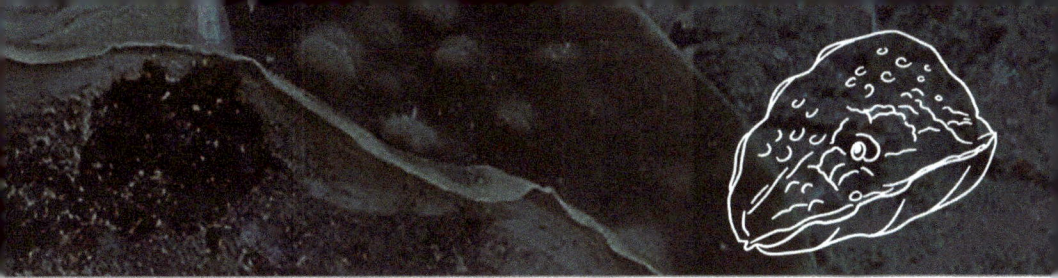

Taxonomy and morphology

The shell is elongate-oval to semicircular, lacks shell ribbing with weak primary radial folds, with a poorly defined byssal orifice. Mantle has *Hippopus*-like features, such as lacking hyaline organs, with little to no mantle extension over the upper shell margins. Although the shell resembles *Tridacna derasa*, this species can be distinguished by its coloured patches on the shell ribbing, rugose mantle surface with numerous protuberances, and incurrent siphon that bears prominent, distinct guard tentacles.

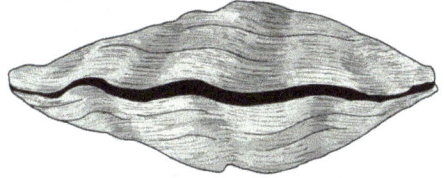

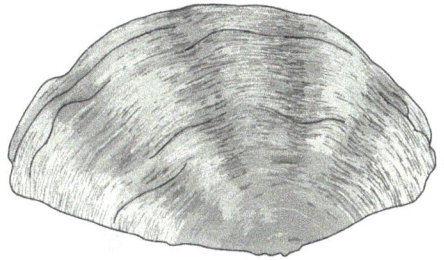

Depth occurrence

20 m

Tonga, Fafa Island

New Caledonia

Close-up of the guard tentacles

Fiji, Fulaga Island (Lau Province)

129

Tridacna derasa
(Röding, 1798)

Tridacna derasa can be found from the Cocos (Keeling) Islands to Tonga and China to Queensland (Australia). It is the second largest giant clam species, which can grow up to 60 cm.

Common name

Smooth Giant Clam

Biology

This species is primarily free-living as adults. It inhabits a wide range of habitats: reef flats, fore reefs, barrier reefs, and atoll lagoons down to depths of 20 m.

Geographic distribution

Taxonomy and morphology

The shell is semicircular to fan-shaped, very thick and heavy, possess almost no ribbing with weak primary radial folds, with a small byssal orifice. Mantle exhibits brilliant colours, displaying shades of blue, orange and green with striped patterns on smooth surfaces. Hyaline organs are scattered near the mantle edge, and the incurrent siphon bears relatively indistinct guard tentacles.

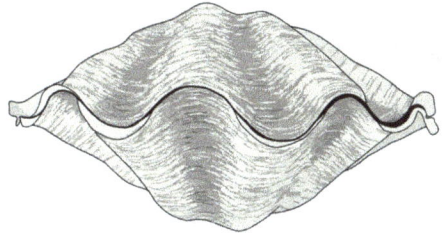

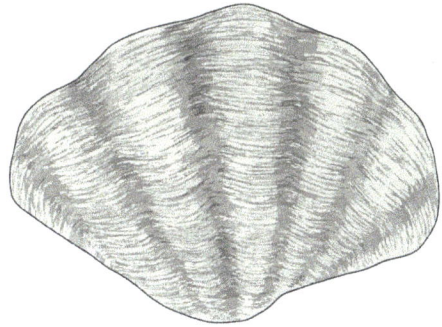

Depth occurrence

20 m

Australia, Great Barrier Reef

New Caledonia

New Caledonia

Australia, Great Barrier Reef

Palau, Kayangel

Philippines, Bolinao

Philippines, Bolinao

133

Tridacna crocea
Lamarck, 1819

Tridacna crocea can be found from Australia to Japan, east to Palau, and Vanuatu to the Andaman Islands. It is the smallest giant clam species, which can grow up to 15 cm.

Common names

Boring Giant Clam, Crocus Clam

Biology

This species is a rock borer that embeds its entire body into the substratum and byssally attaches to its borehole, leaving only the mantle exposed. It mainly inhabits the reef flats in shallow waters of depths no more than 10 m. It appears well adapted to low salinity levels, often found in areas that experience freshwater runoff.

Geographic distribution

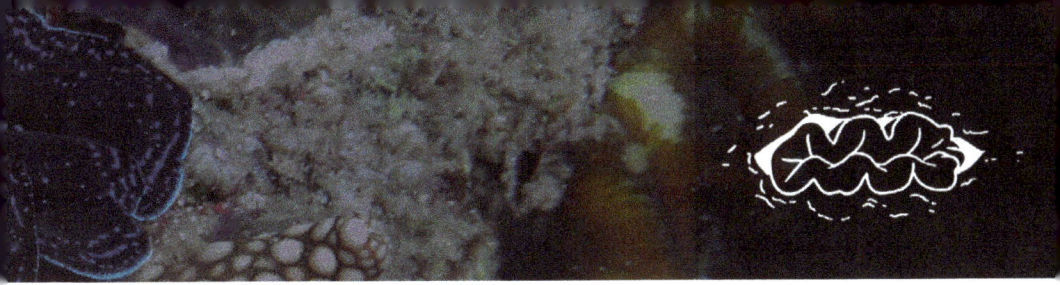

Taxonomy and morphology

The shell is oblique trigonal to oval oblong, thick with very weak primary radial folds, with a wide byssal orifice. This species is usually identified by its boring habit, but it also develops well-spaced scutes that erode over time within the borehole. Mantle is usually brightly coloured, exhibiting shades of blue, green, purple, white and brown. Hyaline organs are numerous across the mantle, and incurrent siphon bears indistinct guard tentacles.

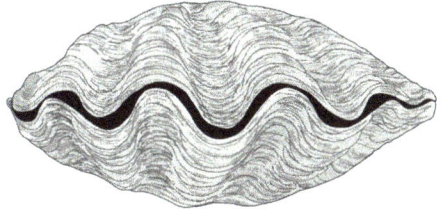

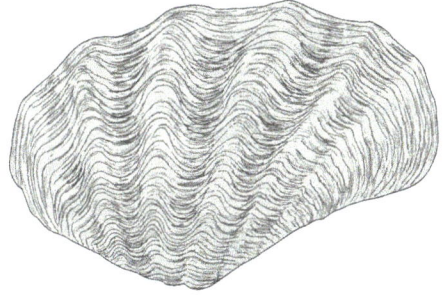

Depth occurrence

10 m

135

Peninsular Malaysia, Tioman Island

Indonesia, Derawan Islands

Thailand, Gulf of Thailand

Palau, Risong

Palau

136

Myanmar, Rakhine, Andrew Bay

Papua New Guinea

Philippines, Escalante

Papua New Guinea

Ryukyu Islands, Okinawa Island

Ryukyu Islands, Okinawa Island

137

Singapore, Pulau Semakau

Singapore, Raffles Lighthouse

138

Singapore, Pulau Semakau

South China Sea, Pulau Layang Layang

Taiwan, Kenting

Tridacna squamosa
Lamarck, 1819

Tridacna squamosa is present from the Red Sea and eastern Africa in the west to the Pitcairn Islands, Ryukyus (southern Japan) and Queensland (Australia) in the east. It can grow to 40 cm long, with the largest specimen recorded at 42.9 cm.

Common names

Fluted Giant Clam, Scaly Giant Clam

Biology

This species may be byssally attached or free-living as adults, while the juvenile is typically byssally attached to the reef substrata. It inhabits a wide depth range, from reef flats to reef slopes down to 42 m, and is usually found in sheltered sites (e.g., wedged between corals).

Geographic distribution

Taxonomy and morphology

The shell is oval-oblong to subtrigonal-oblong, with well-defined ribs and primary radial folds, with a small byssal orifice. Valves are also often coloured (yellow and orange-pink). Scutes are large, semi-circular, and well-spaced out. Mantle usually exhibits mottled patterns in combinations of yellow, orange, blue, green and brown. Hyaline organs are widely spaced along the mantle edge, and the incurrent siphon bears prominent, branched guard tentacles.

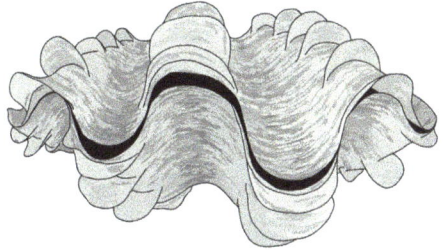

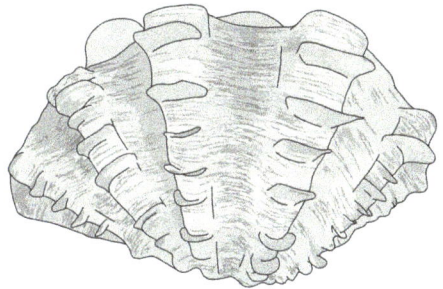

Depth occurrence

42 m

Maldives

Palau, Arch

Mauritius

Papua New Guinea

Mauritius

Papua New Guinea, Madang

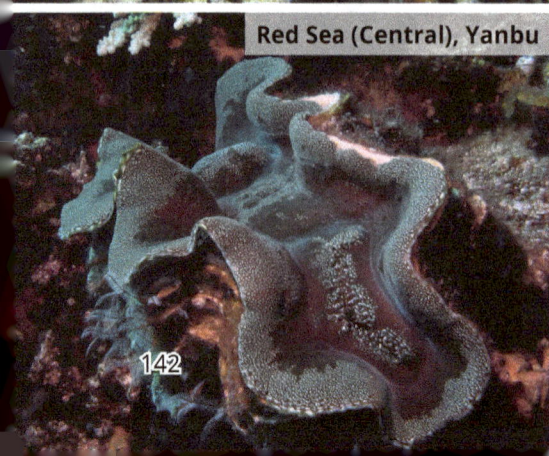
Red Sea (Central), Yanbu

142

Indonesia, Papua

Indonesia, Bintan Island

Thailand, Gulf of Thailand

Indonesia, Bunaken Island

Thailand, Phuket Island

143

East Malaysia, Sabah

Peninsular Malaysia, Perhentian Islands

Peninsular Malaysia, Perhentian Islands

Ryukyu Islands, Okinawa Island

South China Sea, Pulau Layang Layang

144

Singapore, Pulau Semakau

Taiwan, Kenting

145

Tridacna noae
(Röding, 1798)

Tridacna noae is present from the Ryukyus (southern Japan), Taiwan, Southeast Asia, Western Australia, the Pacific Islands, and to the east in Christmas Island. The largest specimen recorded is at 28 cm.

Common names

Noah's Giant Clam, Teardrop Maxima

Biology

As a recently resurrected species, data on its habitat and distribution are scarce, but inferred to be similar to *Tridacna maxima*. This species is often found partially embedded in reef substrata as adults, while juveniles may be fully embedded. It inhabits the shallow reefs and lagoons, occupying depths between 1 and 15 m.

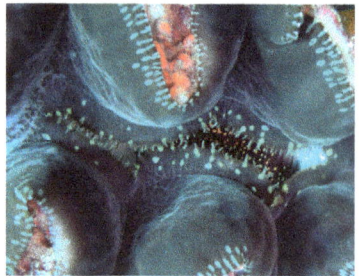

Geographic distribution

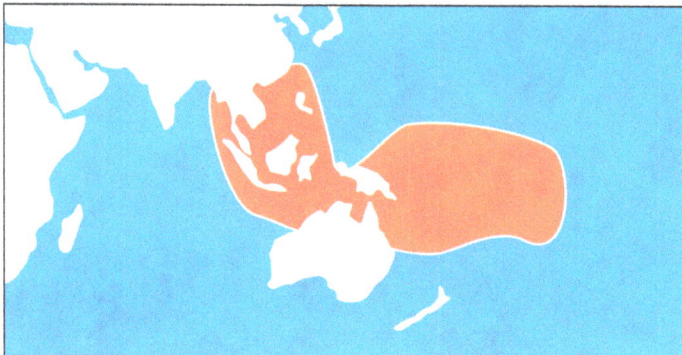

Taxonomy and morphology

The shell is oblique oval to oval-subtrigonal, thick and heavy with 5-7 primary radial folds, with a moderately wide byssal orifice. Scutes are relatively well-spaced out. Mantle exhibits highly distinct ornamentation including discrete teardrop patches typically bounded by white margins, and the surface may possess papillae. Hyaline organs are sparsely distributed and discontinuously arranged along the mantle margin, and the incurrent siphon bears indistinct guard tentacles.

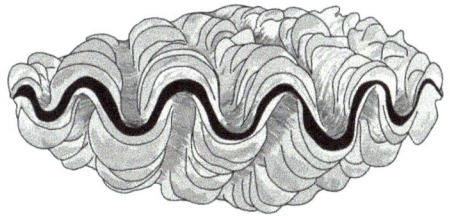

Depth occurrence

15 m

Christmas Island

Indonesia, Kei Island

Myanmar, Rakhine, Andrew Bay

New Caledonia

Papua New Guinea

Ryukyu Islands, Okinawa Island

Papua New Guinea

148

Ryukyu Islands, Okinawa Island

South China Sea, Dongsha Atoll

Taiwan, Kenting

150

Taiwan, Kenting

Close-up of 'tear-drops'

Close-up of the hyaline organs

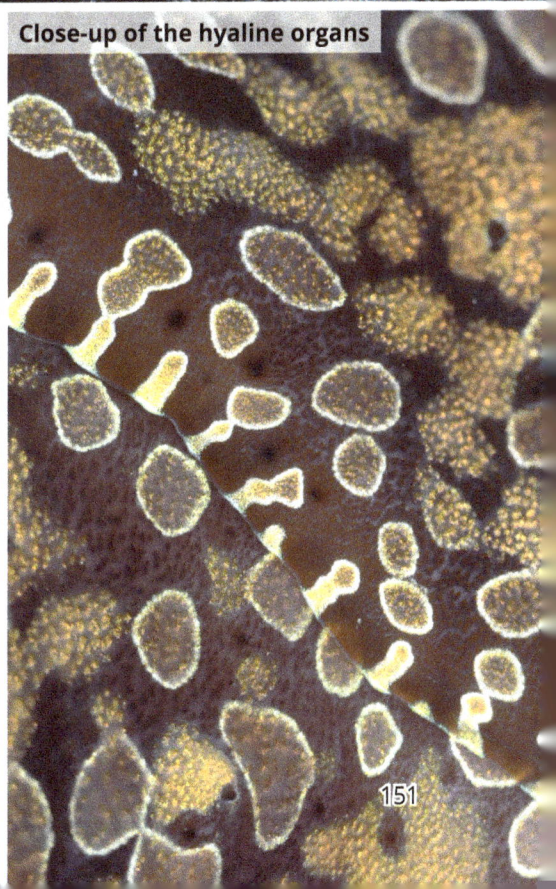

151

Tridacna maxima
(Röding, 1798)

Tridacna maxima is a cosmopolitan species throughout the Indo-Pacific. It can usually grow up to 35 cm long, with the largest specimen recorded at 41.7 cm.

Common name

Small Giant Clam

Biology

Juveniles of this species are usually fully embedded in the reef substrata, but older individuals eventually outgrow the borehole and become partially embedded only. This species tends to byssally attach to the inside of its borehole, as with other boring species. It typically inhabits the shallow areas of reefs and lagoons, rarely beyond a depth of 10 m.

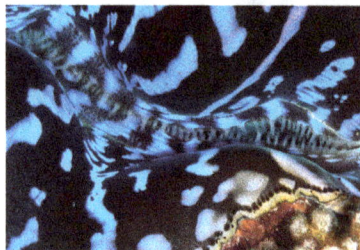

Geographic distribution

Taxonomy and morphology

The shell is oblong rhomboidal to elongate oval, thick and heavy with 4-6 primary radial folds, with a moderately wide byssal orifice. Scutes are relatively close-set together. Mantle is brilliantly coloured (usually blue, green and brown) and mottled. Hyaline organs form a continuous line tightly spaced along its mantle margin, and the incurrent siphon bears indistinct guard tentacles.

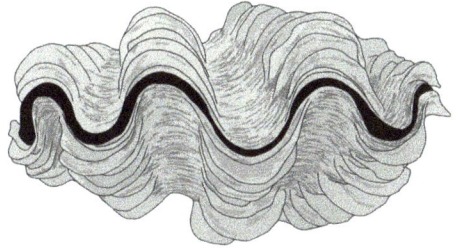

Depth occurrence

10m

Lakshadweep Archipelago

Close-up of the hyaline organs

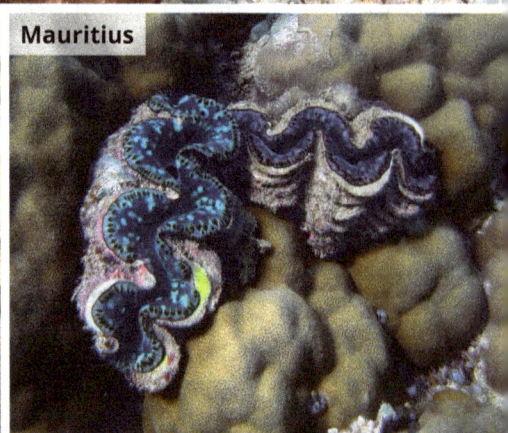
Mauritius

Mauritius

154

Red Sea (Central), Thuwal

Red Sea (Central), Duba

Cook Islands

French Polynesia, Moorea

New Caledonia

Palau, Short

Papua New Guinea

Indonesia, Sangalaki Island

Philippines, Apo Island

Peninsular Malaysia, Perhentian Islands

East Malaysia, Sabah

Peninsular Malaysia, Tioman Island

Myanmar, Rakhine, Andrew Bay

157

Papua New Guinea

South China Sea, Dongsha Atoll

South China Sea, Pulau Layang Layang

Yaeyama Islands, Iriomote Island

Taiwan, Kenting

159

Tridacna squamosina
Sturany, 1899

Tridacna squamosina is currently known from the Red Sea only. It was originally collected during the Pola expedition (1895-1898) to the Red Sea, and recognised by Sturany (1899) as a distinct species: *Tridacna elongata* var. *squamosina*. In the late 2000s, this species was re-introduced as *Tridacna costata*, where the largest individual recorded was 32 cm long. Huber & Eschner (2011) later analysed the materials from the Pola expedition and demonstrated that *Tridacna costata* is a junior synonym of *Tridacna squamosina*.

Biology

This species appears to be weakly byssally attached to the seabed as adults. It occurs in shallow waters (<2 m), including reef flats, seagrass beds, sandy-rubble flats, or under branching corals or coral heads.

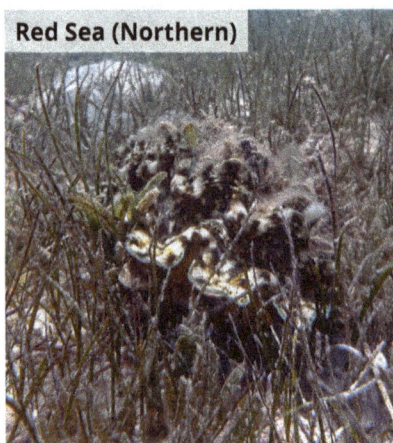

Red Sea (Northern)

Geographic distribution

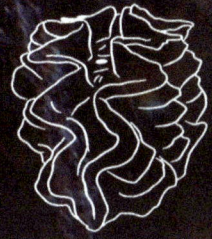

Taxonomy and morphology

The shell is oval-oblong, very thick and heavy with 5-6 strongly projecting and pointed triangular primary radial folds, with a wide byssal orifice. Scutes are large and relatively crowded. Mantle is usually subdued brown, mottled cream with a green margin, and surface is covered with prominent and numerous papillae (knob-like protrusions) in larger specimens. Hyaline organs are scattered on the mantle, and incurrent siphon bears long, branched guard tentacles.

Saudi Arabia

Depth occurrence

2 m

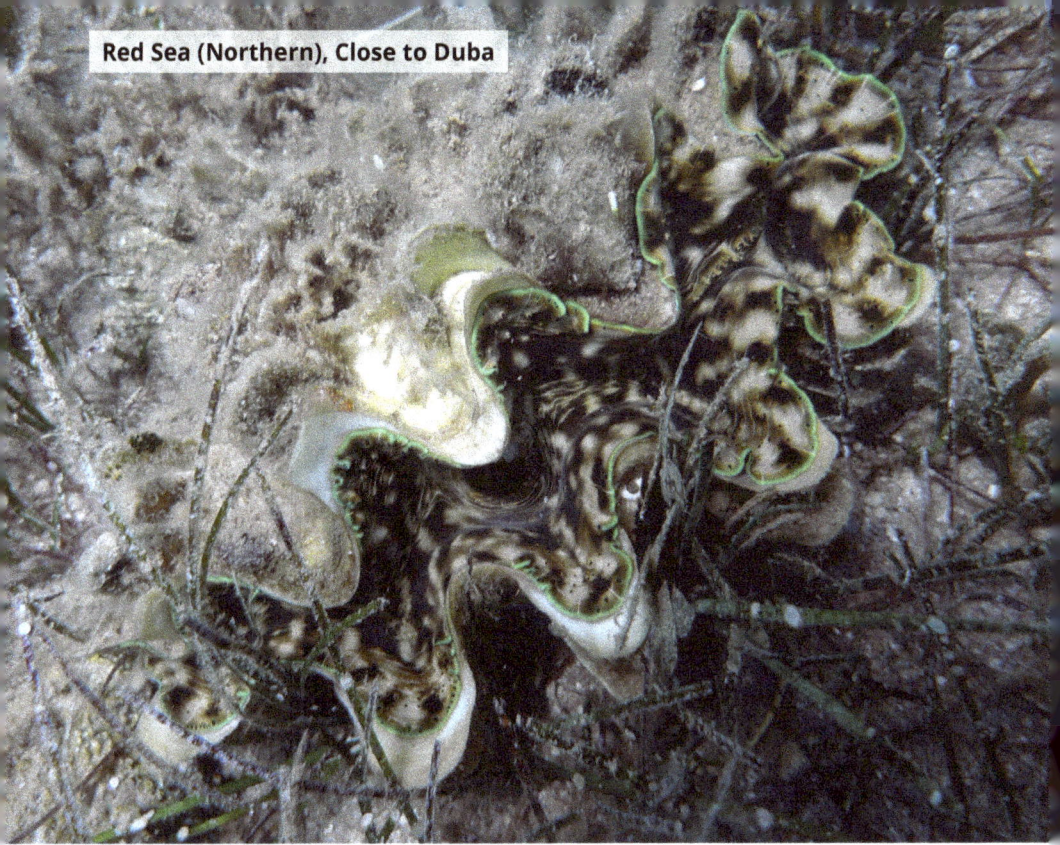
Red Sea (Northern), Close to Duba

Red Sea (Southern), Farasan Banks

163

Tridacna rosewateri
Sirenko & Scarlato, 1991

Tridacna rosewateri is currently only known from the Mauritian waters, specifically in the Mascarene Plateau (Saya de Malha and Nazareth Banks), Cargados Carajos Archipelago (St Brandon) and Tromelin Island. It was first collected from the Saya de Malha Bank, Indian Ocean, during a 1984 expedition and described as a new species by Sirenko & Scarlato (1991). The largest individual collected was 19.1 cm long. In the late 2010s, individuals were re-introduced as *Tridacna lorenzi*, but genetic evidence confirmed that they were, in fact, *Tridacna rosewateri*. Hence, *Tridacna lorenzi* is now considered a junior synonym of *Tridacna rosewateri*. This also proves that *Tridacna rosewateri* has been a valid species since its last sighting in 1984.

Common name
Rosewater's Giant Clam

Biology
This species appears to be partially embedded within reef substrata. Its reported depth range is highly variable, from shallow turbid lagoon waters (0-1 m) to deep reef plateau (12-13 m and 38-39 m).

Geographic distribution

Taxonomy and morphology

The shell is elongate-oval to oval-trigonal, moderately thin walls with 4-5 pronounced primary radial folds, and with a moderately wide byssal orifice. Scutes are larger and more dense on the primary radial folds than *Tridacna squamosa*. Mantle subdued purple-brown. Hyaline organs form a continuous line along the mantle edge, and guard tentacles are indeterminate. The deep triangular valve margin folds distinguish this species from *Tridacna maxima* and *Tridacna squamosa*.

Cargados Carajos Archipelago

Depth occurrence

39 m

Tridacna elongatissima
Bianconi, 1856

Tridacna elongatissima is known from Mozambique, Madagascar, Juan de Nova, Glorieuses, Reunion Island and Mauritius. It was first collected from Mozambique (restricted to Inhambane) and originally recognised by G.G. Bianconi. Although it was previously a junior synonym of *Tridacna maxima* or *Tridacna squamosa*, new morphological and genetic evidence collected in the late 2010s from live individuals revealed it to be distinct and resurrected as a valid species.

Biology

This species is restricted to the reef top (down to 12 m), where individuals live embedded in the reef, usually together with *Tridacna maxima* individuals.

Madagascar

Geographic distribution

Taxonomy and morphology

The shell is oblong trigonal to oblong rhomboidal, heavy and inflated with 6-7 strongly projecting and pointed triangular primary radial folds, with a moderately wide byssal orifice. Scutes are large and close-set together, predominantly covering the upper shell. Mantle exhibits greyish-brown, blue or purple, variably mottled with a green margin. Hyaline organs are scattered on the mantle and mantle edge, and the incurrent siphon bears long, branched guard tentacles.

Juan de Nova

Depth occurrence

12 m

La Réunion

Madagascar, Ste Marie Island

Madagascar, Ste Marie Island

Madagascar, Ste Marie Island

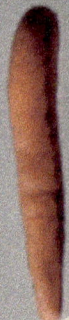

065

069

07

071

077

11. Dichotomous Key

The following dichotomous key is based on shell and mantle characters, with the latter only applying to live specimens. It also includes relevant information on species ecology and geographic occurrence for the uncommon species.

1.   Shell with a very narrow byssal orifice bordered by interlocking teeth. Mantle only reaching to the edge of upper shell margins when fully extended 2

        Shell with a well-defined byssal orifice that does not have interlocking teeth. Mantle, when fully extended, usually projects laterally beyond the shell margins 3

2.   Shell is thick and heavy with strong radial ribbing, reddish blotches in irregular bands crossing prominent radial ribs; with irregular semi-tubular projections on the upper shell margins. Incurrent siphon bears no guard tentacles ....*Hippopus hippopus* (pg. 112)

        Shell is thinner and smoother, usually lacking the reddish blotches; with regularly rounded projections on the upper shell margins. Incurrent siphon bears indistinct guard tentacles ....*Hippopus porcellanus* (pg. 116)

3.   Shell of mature specimens can reach up to 100 cm, with about 4-6 elongate, triangular projections on the upper shell margins; shell symmetry equilateral; shell sculpture without scutes (scale-like projections), except in juvenile specimens. Mantle brownish and olive-green, with numerous hyaline organs bordered by iridescent blue-green circles; incurrent siphon bears no guard tentacles ....*Tridacna gigas* (pg. 122)

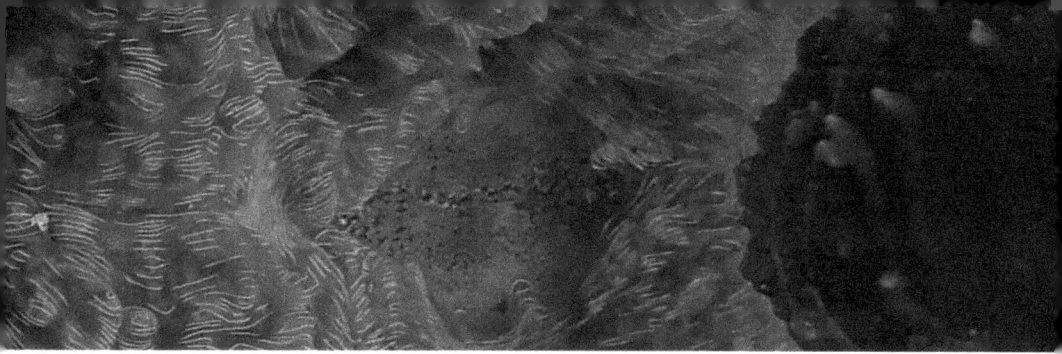

Shell of large specimens can reach up to 60 cm, without elongate triangular projections on the upper shell margins; shell sculpture with or without scutes. Mantle is variably coloured; incurrent siphon bears guard tentacles 4

4.   Shell of specimens usually up to 50 cm, occasionally larger. Mantle colouration is often subdued and mottled, where mantle surface is rugose with protuberances; hyaline organs absent; incurrent siphon bears prominent distinct guard tentacles ....*Tridacna mbalavuana* (pg. 126)

Shell of most specimens usually between 40 to 60 cm. Mantle coloration often bright, mantle surface is usually smooth; hyaline organs present; incurrent siphon bears prominent to indistinct guard tentacles 5

5.   Shell is thick and plain, without scutes (except in juvenile specimens), weak primary radial folds; shell symmetry inequilateral; the hinge is usually longer than half shell length. Mantle often with bright coloration, tending to have elongate patterns of colour, sometimes brilliant blue, mantle surface is smooth; incurrent siphon bears prominent guard tentacles ....*Tridacna derasa* (pg. 130)

Shell has scutes, sometimes with strong primary radial folds; the hinge is equal to or less than half shell length. Mantle colouration may be brightly coloured or subdued, mantle surface is usually smooth 6

6. Shell of specimens about 15 cm; shell symmetry weakly inequilateral; scutes are only evident along shell margins as older scutes on shells are usually worn away; byssal orifice wide. Mantle brightly coloured, often with much colour variation between nearby specimens; hyaline organs numerous across mantle; guard tentacles on incurrent siphon small. Occurs fully embedded up to the shell margins in coral boulders and reef substrata ....*Tridacna crocea* (pg. 134)

Shell of specimens up to about 30-40 cm; shell symmetry variable; scutes present on shells, which may persist in mature individuals; byssal orifice narrow to wide. Occurs free-living on the seabed or partially embedded in the substrata 7

7. Shell typically has 5-6 primary radial folds; shell symmetry approximately equilateral; large and well-spaced scutes; the hinge is about half shell length; byssal orifice small. Mantle subdued colours with mottled patterns; guard tentacles on incurrent siphon large and branching. Occurs free-living on the seabed, may be weakly byssally attached ....*Tridacna squamosa* (pg. 140)

Shell has 4-7 primary radial folds; shell symmetry often strongly inequilateral; the hinge distinctly shorter than half shell length; byssal gape moderately wide to wide 8

8. Shells have broad, moderately projecting, low-rounded primary radial folds; scute spacing well-spaced to close-set; byssal orifice moderately wide. Occurs partially embedded in the substrata 9

Shells have pronounced, strongly projecting, triangular primary radial folds; scute spacing usually well-spaced; byssal orifice moderately wide to wide. Occurs free-living on the seabed, may be weakly byssally attached or partially embedded in the substrata 10

9.   Shell has 5-7 primary radial folds, with scutes relatively well-spaced out. Mantle may exhibit distinct oval patches bounded by white margins; hyaline organs are sparse and discontinuous along the mantle margin; guard tentacles on incurrent siphon small ....*Tridacna noae* (pg. 146)

Shell has 4-6 primary radial folds, with scutes relatively close-set together. Mantle typically has a mottled or dashed pattern; hyaline organs are highly concentrated and continuous along the mantle margin; guard tentacles on incurrent siphon small ....*Tridacna maxima* (pg. 152)

10.   Shell has 5-6 primary radial folds, with large and well-spaced scutes; shell weakly inequilateral; byssal orifice wide. Mantle pattern subdued, with a green margin, mantle surface is covered with numerous pronounced papillae (knob-like protrusions); hyaline organs scattered on the mantle; guard tentacles on incurrent siphon long and branched. Occurs free-living on the seabed, may be weakly byssally attached. Currently, only known from the Red Sea ....*Tridacna squamosina* (pg. 160)

Shell has 4-7 primary radial folds, with scutes relatively well-spaced; shell symmetry weakly inequilateral to inequilateral; byssal orifice moderately wide. Mantle pattern subdued or mottled, mantle surface is smooth. Occurs free-living on the seabed or partially embedded in the substrata. Currently, known from the Western Indian Ocean 11

11.   Shell symmetry weakly inequilateral. Mantle subdued purple-brown; hyaline organs form a continous line along mantle edge. Occur free-living on the seabed; inhabit shallow turbid lagoon waters (0-1 m) to deep reef plateau (down to 39 m) ....*Tridacna rosewateri* (pg. 164)

Shell symmetry inequilateral. Mantle pattern subdued or mottled, with a green margin; hyaline organs scattered on mantle and mantle edge; guard tentacles on incurrent siphon long and branched. Occurs partially embedded in the substrata; restricted to the reef top (down to 12 m) ....*Tridacna elongatissima* (pg. 166)

12. Giant Clams in Vernacular Languages

Across the Indo-Pacific region, the name 'giant clam' may be expressed in the native languages or dialects spoken by people in a particular country or region. These are known as vernacular names given to an organism by the local people, and they are customarily spoken informally rather than written in the native languages. Here, the following table lists the vernacular names that refer to the giant clam or its individual species, compiled to date from various countries in the Indo-Pacific region.

Where the information is available, the wording is either provided in its native language with a pronunciation included in round brackets or the spelling used by native language speakers is adopted. Most names are pronounced approximately as spelt, provided every letter is sounded. When using these vernacular names for reference, caution must be exercised not to use them too rigorously as they may correspond to giant clam sizes rather than a particular species.

Table of Giant Clam Vernacular Names

Country Native language	Common name / Species name	Vernacular name(s)
American Samoa Samoan	Giant clam	Faisua
Cambodia Khmer	Giant clam	គ្រំយក្ស (krom yok)
People's Republic of China Mandarin Chinese (Simplified)	*Hippopus hippopus* *Hippopus porcellanus*	砗蚝 (chē háo) 瓷口砗蚝 (cí kǒu chē háo), 瓷口砗磲 (cí kǒu chē qú), 瓷菱砗磲 (cí líng chē qú)
	Tridacna gigas	库氏砗磲 (kù shì chē qú), 大砗磲 (dà chē qú)
	Tridacna mbalavuana	魔鬼砗磲 (móguǐ chē qú)
	Tridacna derasa	扇砗磲 (shàn chē qú), 无鳞砗磲 (wú lín chē qú)
	Tridacna crocea	番红砗磲 (fān hóng chē qú), 红袍砗磲 (hóng páo chē qú), 圆砗磲 (yuán chē qú)
	Tridacna squamosa	鳞砗磲 (lín chē qú)
	Tridacna maxima	长砗磲 (zhǎng chē qú)
	Tridacna rosewateri	罗氏砗磲 (luōshì chē qú)
Cook Island Maori	Giant clam	Pa'ua
Fiji Bauan Fijian	*Hippopus hippopus* *Tridacna gigas* *Tridacna mbalavuana* *Tridacna derasa* *Tridacna crocea* *Tridacna squamosa* *Tridacna maxima*	Teke ni ose Vasua matau Vasua tevoro Vasua dina Vasua kabi Vasua cega Kata vatu

French Polynesia French	*Tridacna maxima*	Bénitier
French Polynesia Rapa	*Tridacna maxima*	Pagii
French Polynesia Tahitian	Giant clam	Pahua, Kohea
Guam Chamorro	Giant clam	Hima
India (Lakshadweep) Malayalam	Giant clam	Kakka
Indonesia Bahasa Indonesia	*Hippopus hippopus*	Kima Tapak Kuda, Kima Kuku Beruang, Kima Pasir
	Hippopus porcellanus	Kima Cina, Kima Porselen
	Tridacna gigas	Kima Raksasa
	Tridacna derasa	Kima Selatan
	Tridacna crocea	Kima Kunia, Kima Lubang
	Tridacna squamosa	Kima Sisik, Kima Seruling
	Tridacna maxima	Kima Kecil
	Tridacna rosewateri	Kima Mauritius
Indonesia Papuan Malay	*Hippopus hippopus*	Katop arpos
	Hippopus porcellanus	Mangkapdu
	Tridacna gigas	Intef
	Tridacna derasa	Mangkapdu
	Tridacna crocea	Intef
	Tridacna squamosa	Irpur
	Tridacna maxima	Inkonsonsen
Japan (Ryukyu Archipelago) Japanese	*Hippopus hippopus*	シャゴウ (shagou)
	Tridacna gigas	オオシャコガイ (ooshakogai)
	Tridacna derasa	ヒレナシシャコ (hirenashishako)
	Tridacna crocea	ヒメシャコガイ (himeshakogai)
	Tridacna squamosa	ヒレジャコガイ (hirejakogai)
	Tridacna noae	トガリシラナミ (togarishiranami)
	Tridacna maxima	シラナミガイ (shiranamigai)

Kenya Swahili	*Tridacna squamosa* *Tridacna maxima*	Shaza Shaza
West Malaysia Bahasa Melayu	Giant clam	Siput kima, Kima gergasi, Kerang gergasi
East Malaysia Rungus	Giant clam	Kimo
Maldives Dhivehi	Giant clam	Gaahaka
Myanmar Burmese	*Hippopus hippopus* *Tridacna gigas* *Tridacna squamosa* *Tridacna maxima*	Wet Wun Let Thae Ka Mar Kyar Let Thae Jate Ket Pulway Kyar Let Thae Shae Myaw Kyar Let Thae
New Caledonia French	*Hippopus hippopus*	Bénitier rouleur
Niue Niuean	*Tridacna squamosa* *Tridacna maxima*	Gege Gege
Palau Palauan	*Hippopus hippopus* *Hippopus porcellanus* *Tridacna gigas* *Tridacna derasa* *Tridacna crocea* *Tridacna squamosa* *Tridacna maxima*	Duadeb Duadou Oktang Kism Oruer Ribkungal Melibes
Philippines Cebuano	*Hippopus hippopus* *Hippopus porcellanus* *Tridacna gigas* *Tridacna derasa* *Tridacna crocea* *Tridacna squamosa* *Tridacna noae* *Tridacna maxima*	Kukong kabayo Kukong kabayo Tilang dako Tilang dako Let-let Hagdan-hagdan Manlet or Manlot Manlet or Manlot, Sali-ot
Philippines Tagalog	*Tridacna gigas* *Tridacna derasa* *Tridacna crocea* *Tridacna squamosa* *Tridacna maxima*	Dako nga taklobo Hamis nga taklobo Sollot sollot Babaeng taklobo Lalakeng taklobo
Philippines Tausug	Giant clam	Manangkay, Kima

Republic of Kiribati I-kiribati	*Hippopus hippopus* *Tridacna gigas* *Tridacna squamosa* *Tridacna maxima*	Te neitoro Te kima Te were matai Te were
Singapore Malay	*Hippopus hippopus* *Tridacna squamosa*	Siput lupat Siput kima
Solomon Islands **Choiseul** Tavula	*Hippopus hippopus* *Tridacna gigas* *Tridacna derasa* *Tridacna crocea* *Tridacna squamosa* *Tridacna maxima*	Mamasivu Meka Meka Kasiputu Jiku Qeto
Solomon Islands **Guadalcanal** Ghari	*Hippopus hippopus* *Tridacna gigas* *Tridacna crocea* *Tridacna squamosa* *Tridacna maxima*	Kwa-kwa Ghima Kapichi Inuvitasi Kapich
Solomon Islands **Malaita** 'Are'are	*Hippopus hippopus* *Tridacna gigas* *Tridacna derasa* *Tridacna crocea* *Tridacna squamosa* *Tridacna maxima*	Apuri Piawa Sisikeni Unupanu Sisimane Taura
Solomon Islands **New Georgia** Marovo	*Hippopus hippopus* *Tridacna gigas* *Tridacna crocea* *Tridacna squamosa* *Tridacna maxima*	Hohombulu Ose Ulumu Veru-veru Chavi
Solomon Islands **Ranongga** Lungga	*Hippopus hippopus* *Tridacna gigas* *Tridacna derasa* *Tridacna crocea* *Tridacna squamosa* *Tridacna maxima*	Moso Iavo Iavo Gulumu Tatakiru Tatakiru
Solomon Islands **Rennell and Bellona** Rennellese	*Hippopus hippopus* *Tridacna gigas* *Tridacna derasa* *Tridacna crocea* *Tridacna squamosa* *Tridacna maxima*	Takamou Langinga Langinga Ghunu Hāsua Hāsua

Solomon Islands Santa Isabel Bughotu	*Hippopus hippopus* *Tridacna gigas* *Tridacna crocea* *Tridacna squamosa* *Tridacna maxima*	Sepila Tungi Kaspot Fafalche Tunuga
Solomon Islands Sikaiana Sikaiana	*Hippopus hippopus* *Tridacna gigas* *Tridacna derasa* *Tridacna crocea* *Tridacna squamosa* *Tridacna maxima*	Te-pasua Te-tane Te-tane Te-kunu Te-kete hatu Te-veni veni
Solomon Islands Vella Lavella Bilua	*Hippopus hippopus* *Tridacna gigas* *Tridacna derasa* *Tridacna crocea* *Tridacna squamosa* *Tridacna maxima*	Moso Siavo Siavo Tupi-tupi Tataikiri Tataikiri
Republic of China, Taiwan Mandarin Chinese (Traditional)	*Hippopus hippopus* *Hippopus porcellanus* *Tridacna gigas* *Tridacna mbalavuana* *Tridacna derasa* *Tridacna crocea* *Tridacna squamosa* *Tridacna maxima*	硨蚝 (chē háo) 瓷口硨蚝 (cí kǒu chē háo), 菱硨磲蛤 (líng chē qú há), 瓷硨磲蛤 (cí chē qú há) 庫氏硨磲 (kù shì chē qú) 鬼硨磲蛤 (guǐ chē qú há) 扇硨磲蛤 (shàn chē qú há) 番紅硨磲 (fān hóng chē qú) 鱗硨磲蛤 (lín chē qú há) 長硨磲蛤 (zhǎng chē qú há)
Thailand Thai	Giant clam	หอยมือเสือ (ȟxy mʉ̌xšeʉ̌x)
Tuvalu Tuvaluan	Giant clam	Fasua
Viet Nam Vietnamese	Giant clam	Trai tai tượng khổng lồ, Ốc tai bò, Ốc tai tượng
Western Samoa Samoan	*Tridacna squamosa* *Tridacna maxima*	Faisua Faisua

^Author's disclaimer: This table is by no means exhaustive!

Coastal communities living in East Nusa Tenggara, the southernmost province of Indonesia, often use giant clam shells as containers for making sea salt. These shells are filled with seawater and allowed to evaporate in the open air, producing sea salt for local use and consumption. The coastal community in Keliha Beach, Sabu Raijua Regency, is one such community that practises this tradition.

Acknowledgements

The idea for this field guide came to me as a 'Eureka' moment after seeing the beautiful illustrations of giant clam shells that my student, Benjamin Leow, had worked hard to create for his final year project. His creations inspired me to conceptualise what this field guide could look like and how my repository of writings about the giant clams over the years comes in handy in building up the book's content. Thank you very much, Ben, for planting this seed that kept growing in my head!

To my colleagues. Special thanks to Christina Choy and Samuel Lee for reading the early drafts, giving me advice on the illustrations and content, and taking time out of their schedules to pore over the details of the book layout; Bee Yan Lee for co-creating with me an accurate taxonomic key for species identification; Theresa Su for giving me optimism when I felt lost in this book project; and the extraordinary Adel Lee for illustrating and designing this entire field guide. These individuals were as crucial to getting this book done as I was, and my heartfelt gratitude to them for accompanying me on this journey.

To my family. My heartfelt gratitude to my partner, Alex, for being my biggest supporter over the years. And to our dearest daughters, Ada and Maya, for inspiring me daily to keep doing good for the oceans.

To the many others who have helped me so much. Thanks to Rhaimie Wahap and Ruth Wan for introducing me to the world of book publishing; Joy Quek for handling my book from start to end; Nathan Fedrizzi and Michele Haynes (Pew Marine Fellows Program) for helping me navigate the fellowship; Jan Johan ter Poorten (Field Museum of Natural History), Richard Braley (Aquasearch Aquarium), Nicholas Yap (Tropical Marine Science Institute) and Audrey Tan (Centre for Nature-based Climate Solutions/Tropical Marine Science Institute) for graciously taking time to provide feedback and reviews for this field guide.

To all the photo contributors who answered my call for help. This field guide has come to life with all of your beautiful images of giant clams found across the Indo-Pacific. Thanks to everyone for allowing me to use and share your pictures here: Deepak Apte (Bombay Natural History Society); Thierry Baboulenne (Babou Côte Océan); Andrew W. Bruckner (Coral Reef CPR Charity and Consultancy); Roger G. Dolorosa (Western Philippines University - Puerto Princesa Campus); Cécile Fauvelot (Institut de Recherche pour le Développement); James Fatherree (Hillsborough Community College); Bee Yan Lee (Tropical Marine Science Institute/Lee Kong Chian Natural History Museum); Li Keat Lee (Universiti Malaya); Samuel Lee (National University of Singapore); Rahul Mehrotra (Aow Thai Marine Ecology Center); Thane A. Militz (University of Sunshine Coast); Naung Naung Oo (Mawlamyine University); Gustav Paulay (Florida Museum of Natural History); Steven W. Purcell (Southern Cross University); Sundy Ramah (University of Mauritius); Susann Roßbach (The Red Sea Development Company); Julia R. Tapilatu (University of Papua); Heok Hui Tan (Lee Kong Chian Natural History Museum); Victor Tang (Wodepigu Pixels); Teddy Triandiza (Indonesian Institute of Sciences); Charles Waters (Marine Science Consulting).

Finally, an extra special thank you to my 2021 Pew Marine Fellows: Amanda Bates, Rachel Graham, Gakushi Ishimura, Kristen Oleson, Tries Blandine Razak, Yunne-Jai Shin, Rick Stuart-Smith and Songlin Wang. Your ocean optimism encouraged me to do my utmost best in marine conservation.

The development of this book was generously supported by a 2021 Pew Fellowship in Marine Conservation awarded to Mei Lin Neo.

Glossary

A brief dictionary of terminology used.

Adductor muscles are the primary muscle system in bivalve molluscs, containing both smooth and striated fibres between the two valves to hold them closed.

Autotrophic refers to requiring only carbon dioxide or carbonates as a source of carbon and a simple inorganic nitrogen compound for the metabolic synthesis of organic molecules (such as glucose).

Bivalve is a type of mollusc with two shell valves joined by a hinge.

Broadcast spawning is one mode of reproduction in the sea that involves the release of eggs and sperm into the environment, where gametes make contact and lead to external fertilisation, usually without parental care.

Broodstock refers to a group of sexually mature individuals (of both sexes) used in aquaculture for breeding purposes. These organisms are kept in captivity as a source of renewal for, or enhancement of, seed and fry numbers.

Byssus is a structure unique to bivalves, which refers to a set of elastic or calcified filaments secreted by a byssal gland in the foot that anchors the bivalve to a hard surface. **Byssal** is related to the byssus.

Commensalism (in biology) is a long-term relationship between individuals of two species in which one species obtains food or benefits from the other without harming or benefiting the latter. An organism in this relationship is called a **commensal**.

Conspecific (in biology) is a term used to describe individuals belonging to the same species level.

Ctenidium (in molluscs) is the gills that typically consist of respiratory structures that resemble a comb or feather, which perform several functions, including as a site for gaseous exchange and a major site for gathering and sorting food.

Dinoflagellates are motile unicellular algae characterised by a pair of flagella. Most dinoflagellates are photosynthetic, while others are mixotrophic.

Dorsal (in anatomy) is on or relating to the upper side or back of an animal, plant, or organ. In the case of a bivalve, this refers to the umbo or hinge area, where the valves join together.

Ecosystem engineers are species that can create or modify their environment significantly and, in the process, regulate the availability of resources for other species through their presence.

Endosymbionts are organisms that form a symbiotic relationship with another cell or organism.

Epibenthic (in ecology) refers to organisms that live on or just above the bottom sediments in a body of water.

Escape size (in aquaculture) is the size at which an organism becomes less vulnerable to predation.

Excurrent refers to an outgoing current of a vessel or opening.

Fecundity is the potential for reproduction of an organism or a population, usually measured by the number of gametes, seed set or asexual propagules and the survival of the young.

Flagship species (in conservation biology) are chosen to raise support for biodiversity conservation in a given place or social context.

Functional diversity (in biodiversity) measures the value and range of functional traits that organisms contribute to the functioning of communities and ecosystems.

Gamete is a reproductive cell of an animal or plant. In animals, the female gametes are called ova or egg cells, and the male gametes are called sperm.

Gonad is the primary reproductive organ that produces reproductive cells (or gametes). In animals, the female gonad is called the ovary, and the male gonad is called the testis.

Heterotrophic refers to requiring complex organic compounds of nitrogen and carbon (such as that obtained from plant or animal matter) for metabolic synthesis.

Hermaphrodite is an organism with both male and female sex organs or other sexual characteristics and, therefore, can produce both gametes associated with the male and female sexes.

Hinge ligament is made of an elastic protein that connects the two valves along the hinge margin; it is mostly brown or black and serves to open the shell when the adductor muscles relax.

Holobiont is a complex assemblage of organisms that includes a multicellular host and its associated microorganisms. Hence, giant clams are holobionts that include the clam itself, photosynthetic dinoflagellates Symbiodiniaceae, and associated bacteria and viruses.

Hyaline organs (in giant clams) refer to the small dark pinhole eyes usually found on the margins of the mantle.

Incurrent refers to an incoming current of a vessel or opening.

Inequivalve is a condition where the shell valves are unequal in size.

Iridocyte (also called iridophore) is a specialised cell containing iridescent crystals of guanine found in some animals, such as reptiles and cephalopods, giving these animals their iridescence.

Mantle (in molluscs) is a significant part of the anatomy of molluscs: it is the dorsal body wall which covers the visceral mass (i.e., digestive, nervous, excretory, reproductive and respiratory systems) and usually protrudes in the form of flaps beyond the main body, and secretes the shell.

Mariculture is a specialised sector of aquaculture that focuses on farming of marine plants and animals for food, medicine, or any industrial applications.

Metamorphosis is the biological process of transformation from an immature form to an adult form in two or more distinct life stages.

Monophyletic is a term to describe a group of organisms that are classified in the same taxon and share a most common recent ancestor. A monophyletic group includes all descendents of that most common recent ancestor.

Mixotrophic is the condition of obtaining nourishment from both autotrophic and heterotrophic mechanisms.

Oogenesis is the growth process in which the primary egg cell (or ovum) becomes a mature ovum.

Pediveliger is a late veliger that can use its foot to crawl and provide temporary attachment to the seabed.

Pelagic refers to living or occurring in the open sea.

Photosynthesis is a process by which phototrophs use sunlight, water and carbon dioxide to generate oxygen and energy in the form of sugar.

Photosymbiosis is a symbiotic relationship between a heterotrophic host with one or more types of photosynthetic microalgae.

Phototrophy refers to a metabolic process in which energy from the sun is captured and converted into chemical energy for growth.

Phylogeny refers to the history of the evolution of a species or group, especially concerning the lines of descent and relationships among broad groups of organisms.

Phylogenetic diversity (in biodiversity) measures biodiversity based on the evolutionary history among species in a given community that can be computed using taxonomy or phylogeny.

Radial fold is a type of external sculpture and texture of a bivalve shell. It refers to any linear pattern originating at the beak or umbo and radiating towards the shell margins.

Riblet is a type of raised radial sculpture for radial patterns on shells, and it refers to small ribs that form well-defined, narrow ridges on the shell texture.

Satoumi is a Japanese term describing a desirable state of coastal and marine zones with enhanced biodiversity and productivity realised by active human interventions.

Scutes are produced extensions of external ribs projecting outward from the external shell surface.

Siphon is a tubular organ in an aquatic animal, especially a mollusc, used to convey water in or out of the mantle cavity.

Species richness is the number of species represented in an ecological community, landscape or region.

Symbiosis is a long-term biological interaction between two organisms in a close physical association. These associations may be positive (mutualism, commensalism) and negative (parasitism), and the members are called **symbionts**. **Symbiont** is an organism living in symbiosis with another.

Spermatogenesis is the production and development of mature spermatozoa.

Taxonomic diversity (in biodiversity) is a principal measure of biodiversity, inferred based on the number of species present in a specific locality of a region.

Trochophore is a small, translucent, free-swimming planktonic marine larva characteristic of most groups of molluscs. It is spherical or pear-shaped and encircled by the prototroch (a ciliary ring consisting of minute hair-like structures), enabling it to swim.

Umbo (in zoology) is a protuberance at the highest point of each shell valve of a bivalve mollusc.

Velum (in zoology) is a ciliated organ used for swimming, gas exchange, and food collection. It is usually present in molluscs larvae.

Veliger is the planktonic larva stage of certain molluscs, which typically bears two ciliated paddles (velum) for swimming and feeding.

Ventral (in anatomy) is on or relating to the underside of an animal or plant. In the case of a bivalve, this usually refers to the region of its foot.

Zooxanthellae is a colloquial term for single-celled dinoflagellates that can live in symbiosis with diverse marine invertebrates.

References

1. INTRODUCTION

Lucas, J.S. (1988). Giant clams: description, distribution and life history. In: J.W. Copland & J.S. Lucas (eds). *Giant clams in Asia and the Pacific* (pp. 21–32). ACIAR Monograph No. 9.

Neo, M.L., Wabnitz, C.C.C., Braley, R.D., Heslinga, G.A., Fauvelot, C., Van Wynsberge, S., et al. (2017). Giant Clams (Bivalvia: Cardiidae: Tridacninae): A Comprehensive Update of Species and Their Distribution, Current Threats and Conservation Status. *Oceanography and Marine Biology: An Annual Review*, 55, 87–388.

Spalding, M.D., Fox, H.E., Allen, G.R., Davidson, N., Ferdaña, Z.A., Finlayson, M., et al. (2007). Marine Ecoregions of the World: A Bioregionalization of Coastal and Shelf Areas. *BioScience*, 57(7), 573–583.

Tan, E.Y.W., Neo, M.L. & Huang, D. (2022). Assessing taxonomic, functional and phylogenetic diversity of giant clams across the Indo-Pacific for conservation prioritization. *Diversity and Distributions*, 28(10), 2124–2138.

Yamaguchi, M. (1977). Conservation and cultivation of giant clams in the Tropical Pacific. *Biological Conservation*, 11, 13–20.

2. TAXONOMY AND SYSTEMATICS

Borsa, P., Fauvelot, C., Tiavouane, J., Grulois, D., Wabnitz, C., Abdon Naguit, M.R., et al. (2015). Distribution of Noah's giant clam, *Tridacna noae*. *Marine Biodiversity*, 45(2), 339–344.

Fauvelot, C., Zuccon, D., Borsa, P., Grulois, D., Magalon, H., Riquet, F., et al. (2020). Phylogeographical patterns and a cryptic species provide new insights into Western Indian Ocean giant clams phylogenetic relationships and colonization history. *Journal of Biogeography*, 47(5), 1086–1105.

Huber, M. & Eschner, A. (2010). *Tridacna (Chametrachea) costata* Roa-Quiaoit, Kochzius, Jantzen, Al-Zibdah & Richter from the Red Sea, a junior synonym of *Tridacna squamosina* Sturany, 1899 (Bivalvia, Tridacnidae). *Annalen des Naturhistorischen Museums in Wien. Serie B für Botanik und Zoologie*, 112, 153–162.

Lucas, J.S., Ledua, E. & Braley, R.D. (1991). *Tridacna tevoroa* Lucas, Ledua and Braley: a recently-described species of giant clam (Bivalvia; Tridacnidae) from Fiji and Tonga. *The Nautilus*, 105(3), 92–103.

Mikkelsen, P.M. & Bieler, R. (2008). *Seashells of Southern Florida. Living Marine Mollusks of the Florida Keys and Adjacent Regions: Bivalves*. Princeton University Press.

Monsecour, K. (2016). A New Species of Giant Clam (Bivalvia: Cardiidae) from the Western Indian Ocean. *Conchylia*, 46, 69–77.

Newman, W.A. & Gomez, E.D. (2000). On the status of giant clams, relics of Tethys (Mollusca: Bivalvia: Tridacninae). In: M. Moosa et al. (eds). *Proceedings of the 9th International Coral Reef Symposium, Bali, Indonesia, 23–27 October 2000, Volume 2 (pp. 927–936)*. Jakarta: Indonesian Institute of Sciences, Jakarta: Ministry of Environment, Honolulu, Hawaii: International Society for Reef Studies.

Richter, C., Roa-Quiaoit, H., Jantzen, C., Al-Zibdah, M. & Kochzius, M. (2008). Collapse of a New Living Species of Giant Clam in the Red Sea. *Current Biology*, 18(17), 1349–1354.

Rosewater, J. (1965). The family Tridacnidae in the Indo-Pacific. *Indo-Pacific Mollusca*, 1, 7–82.

Rosewater, J. (1982). A new species of *Hippopus* (Bivalvia: Tridacnidae). *The Nautilus*, 96, 3–6.

Schneider, J.A. (1998). Phylogeny of the Cardiidae (Bivalvia): Phylogenetic Relationships and Morphological Evolution within the Subfamilies Clinocardiinae, Lymnocardiinae, Fraginae and Tridacninae. *Malacologia*, 40(1-2), 321–373

Schneider, J.A. & Ó Foighil, D. (1999). Phylogeny of Giant Clams (Cardiidae: Tridacninae) Based on Partial Mitochondrial 16 rDNS Gene Sequences. *Molecular Phylogenetics and Evolution*, 13(1), 59–66.

Sirenko, B.I. & Scarlato, O.A. (1991). *Tridacna rosewateri* sp. n. A new species of giant clam from the Indian Ocean. *La Conchiglia*, 22, 4–9.

Su, Y., Hung, J.-H., Kubo, H. & Liu, L.-L. (2014). *Tridacna noae* (Röding, 1798) - a valid giant clam species separated from *T. maxima* (Röding, 1798) by morphological and genetic data. *Raffles Bulletin of Zoology*, 62, 124–135.

Tan, E.Y.W., Quek, Z.B.R., Neo, M.L., Fauvelot, C. & Huang, D. (2022). Genome skimming resolves the giant clam (Bivalvia: Cardiidae: Tridacninae) tree of life. *Coral Reefs*, 41(3), 497–510.

3. BIOLOGY

Aharon, P. (1991). Recorders of reef environment histories: stable isotopes in corals, giant clams, and calcareous algae. *Coral Reefs*, 10(2), 71–90.

Aubert, A., Lazareth, C.E., Cabioch, G., Boucher, H., Yamada, T., Iryu, Y., et al. (2009). The tropical giant clam *Hippopus hippopus* shell, a new archive of environmental conditions as revealed by sclerochronological and δ18O profiles. *Coral Reefs*, 28(4), 989–998.

Belda-Baillie, C.A., Sison, M., Silvestre, V., Villamor, K., Monje, V., Gomez, E.D., et al. (1999). Evidence for changing symbiotic algae in juvenile tridacnids. *Journal of Experimental Marine Biology and Ecology*, 241, 207–221.

Calumpong, H.P. (ed.) (1992). *The Giant Clam: an ocean culture manual*. ACIAR Monograph No. 16.

DeBoer, T.S., Baker, A.C., Erdmann, M.V., Ambariyanto, Jones, P.R. & Barber, P.H. (2012). Patterns of *Symbiodinium* distribution in three giant clam species across the biodiverse Bird's Head region of Indonesia. *Marine Ecology Progress Series*, 444, 117–132.

Dumas, P., Tiavouane, J., Senia, J., Willam, A., Dick, L. & Fauvelot, C. (2014). Evidence of early chemotaxis contributing to active habitat selection by the sessile giant clam *Tridacna maxima*. *Journal of Experimental Marine Biology and Ecology*, 452, 63–69.

Griffiths, D.J., Winsor, H. & Luong-Van, T. (1992). Iridophores in the mantle of giant clams. *Australian Journal of Zoology*, 40, 319–326.

Hill, R.W., Armstrong, E.J., Inaba, K., Morita, M., Tresguerres, M., Stillman, J.H., et al. (2018). Acid secretion by the boring organ of the burrowing giant clam, *Tridacna crocea*. *Biology Letters*, 14(6), 20180047.

Holt, A.L., Vahidinia, S., Gagnon, Y.L., Morse, D.E. & Sweeney, A.M. (2014). Photosymbiotic giant clams are transformers of solar flux. *Journal of The Royal Society Interface*, 11(101), 20140678.

Huang, D., Todd, P.A. & Guest, J.R. (2007). Movement and aggregation in the fluted giant clam (*Tridacna squamosa* L.). *Journal of Experimental Marine Biology and Ecology*, 342(2), 269–281.

Ikeda, S., Yamashita, H., Kondo, S., Inoue, K., Morishima, S. & Koike, K. (2017). Zooxanthellal genetic varieties in giant clams are partially determined by species-intrinsic and growth-related characteristics. *PLOS ONE*, 12(2), e0172285.

Jorissen, H., Galand, P.E., Bonnard, I., Meiling, S., Raviglione, D., Meistertzheim, A.-L., et al. (2021). Coral larval settlement preferences linked to crustose coralline algae with distinct chemical and microbial signatures. *Scientific Reports*, 11(1), 14610.

Kamishima, Y. (1990). Organization and development of reflecting plates in iridophores of the giant clam, *Tridacna crocea* Lamarck. *Zoological Science*, 7, 63–72.

Kawaguti, S. & Mabuchi, K. (1969). Electron microscopy on the eyes of the giant clam. *Biological Journal of Okayama University*, 15, 87–100.

Kirkendale, L. & Paulay, G. (2017). "Treatise Online no. 89: Part N, Revised, Volume 1, Chapter 9: Photosymbiosis in Bivalvia". *Treatise Online*.

Klumpp, D.W. & Griffiths, C.L. (1994). Contributions of phototrophic and heterotrophic nutrition to the metabolic and growth requirements of four species of giant clam (Tridacnidae). *Marine Ecology Progress Series*, 115, 103–115.

Klumpp, D.W. & Lucas, J.S. (1994). Nutritional ecology of the giant clams *Tridacna tevoroa* and *T. derasa* from Tonga: influence of light on filter-feeding and photosynthesis. *Marine Ecology Progress Series*, 107, 147–156.

Klumpp, D.W., Bayne, B.L. & Hawkins, A.J.S. (1992). Nutrition of the giant clam *Tridacna gigas* (L.) I. Contribution of filter feeding and photosynthates to respiration and growth. *Journal of Experimental Marine Biology and Ecology*, 155(1), 105–122.

LaJeunesse, T.C., Parkinson, J.E., Gabrielson, P.W., Jeong, H.J., Reimer, J.D., Voolstra, C.R., et al. (2018). Systematic Revision of Symbiodiniaceae Highlights the Antiquity and Diversity of Coral Endosymbionts. *Current Biology*, 28(16), 2570–2580.

Leggat, W., Rees, T.A.V. & Yellowlees, D. (2000). Meeting the photosynthetic demand for inorganic carbon in an alga-invertebrate association: preferential use of CO_2 by symbionts in the giant clam *Tridacna gigas*. *Proceedings of the Royal Society B*, 267, 523–529.

Lim, S.S.Q., Huang, D., Soong, K. & Neo, M.L. (2019). Diversity of endosymbiotic Symbiodiniaceae in giant clams at Dongsha Atoll, northern South China Sea. *Symbiosis*, 78(3), 251–262.

Lucas, J.S. (1994). The biology, exploitation, and mariculture of giant clams (Tridacnidae). *Reviews in Fisheries Science*, 2(3), 181–223.

Munro, P.E., Beard, J.H. & Lacanienta, E. (1983). Investigations on the substance which causes sperm release in tridacnid clams. *Comparative Biochemistry and Physiology*, 74C(1), 219–223.

Neo, M.L., Todd, P.A., Teo, S.L.-M. & Chou, L.M. (2009). Can artificial substrates enriched with crustose coralline algae enhance larval settlement and recruitment in the fluted giant clam (*Tridacna squamosa*)? *Hydrobiologia*, 625(1), 83–90.

Neo, M.L., Vicentuan, K., Teo, S.L.-M., Erftemeijer, P.L.A. & Todd, P.A. (2015). Larval ecology of the fluted giant clam, *Tridacna squamosa*, and its potential effects on dispersal models. *Journal of Experimental Marine Biology and Ecology*, 469, 76–82.

Norton, J.H. & Jones, G.W. (1992). *The Giant Clam: an anatomical and histological atlas*. ACIAR Monograph.

Norton, J.H., Shepherd, M.A., Long, H.M. & Fitt, W.K. (1992). The Zooxanthellal Tubular System in the Giant Clam. *The Biological Bulletin*, 183(3), 503–506.

Rossbach, S., Subedi, R.C., Ng, T.K., Ooi, B.S. & Duarte, C.M. (2020). Iridocytes mediate photonic cooperation between giant clams (Tridacninae) and their photosynthetic symbionts. *Frontiers in Marine Science*, 7, 465.

Sim, D.Z.H., Neo, M.L., Ang, A.C.F., Ying, L.S.M. & Todd, P.A. (2018). Trade-offs between defence and competition in gregarious juvenile fluted giant clams (*Tridacna squamosa* L.). *Marine Biology*, 165(6), 103.

Soo, P. & Todd, P.A. (2014). The behaviour of giant clams (Bivalvia: Cardiidae: Tridacninae). *Marine Biology*, 161(12), 2699–2717.

Yonge, C.M. (1936). Mode of life, feeding, digestion and symbiosis with zooxanthellae in the Tridacnidae. *Scientific Reports, Great Barrier Reef Expedition 1928-29*, 1, 283–321.

4. ECOLOGICAL ROLES IN CORAL REEFS

Arrosa, S., Martin, C., Rossbach, S. & Duarte, C. (2019). Microplastic removal by Red Sea giant clam (*Tridacna maxima*). *Environmental Pollution*, 252, 1257–1266.

Blanco, G.J. & Ablan, G.L. (1939). A rare parasitic crab new to Pangasinan Province, Luzon. *Philippine Journal of Science*, 70(2), 217–219.

Bruce, A.J. (2000). Biological observations on the commensal shrimp *Paranchistus armatus* (H. Milne Edwards) (Crustacea: Decapoda: Pontoniinae). *The Beagle: Records of the Museums and Art Galleries of the Northern Territory*, 16, 91–96.

Cabaitan, P.C., Gomez, E.D. & Aliño, P.M. (2008). Effects of coral transplantation and giant clam restocking on the structure of fish communities on degraded patch reefs. *Journal of Experimental Marine Biology and Ecology*, 357(1), 85–98.

Denton, G.R.W. & Winsor, L. (1986). Giant clams as pollution indicators. *Oceanus*, 29(2), 63.

Duquesne, S.J. & Coll, J.C. (1995). Metal accumulation in the clam *Tridacna crocea* under natural and experimental conditions. *Aquatic Toxicology*, 32(2), 239–253.

Duquesne, S., Flowers, A.E. & Coll, J.C. (1995). Preliminary evidence for a metallothionein-like heavy metal-binding protein in the tropical marine bivalve *Tridacna crocea*. *Comparative Biochemistry and Physiology Part C: Pharmacology, Toxicology and Endocrinology*, 112(1), 69–78.

Fankboner, P.V. (1971). Intracellular Digestion of Symbiontic Zooxanthellae by Host Amoebocytes in Giant Clams (Bivalvia: Tridacnidae), with a Note on the Nutritional Role of the Hypertrophied Siphonal Epidermis. *Biological Bulletin*, 141(2), 222–234.

Fujino, T. (1975). Fine Features of the Dactylus of the Ambulatory Pereiopods in a Bivalve-Associated Shrimp, *Anchistus miersi* (De Man), under the Scanning Electron Microscope (Decapoda, Natantia, Pontoniinae). *Crustaceana*, 29(3), 252–254.

De Grave, S. (1999). Pontoniinae (Crustacea: Palaemonidae) associated with bivalve molluscs from Hansa Bay, Papua New Guinea. *Bulletin de L'Institut Royal des Sciences Naturelles de Belgique - Bulletin van het Koninklijk Belgisch Instituut voor Natuurwetenschappen*, 69, 125–141.

Gutiérrez, J.L., Jones, C.G., Strayer, D.L. & Iribarne, O.O. (2003). Mollusks as ecosystem engineers: the role of shell production in aquatic habitats. *Oikos*, 101(1), 79–90.

Ishii, T., Okoshi, K., Otake, T. & Nakahara, M. (1992). Concentrations of elements in tissues of four species of Tridacnidae. *Nippon Suisan Gakkaishi*, 58(7), 1285–1290.

Itoh, A., Kabe, N., Kuwae, S., Oura, E., Hisamatsu, S., Nakano, Y., et al. (2017). Multi-Element Profiling Analyses of Symbiotic Zooxanthellae and Soft Tissues in a Giant Clam (*Tridacna crocea*) Living in the Coral Reefs and Their Intake Process of Zn and Cd. *Bulletin of the Chemical Society of Japan*, 90(5), 520–526.

Maboloc, E.A. & Mingoa-Licuanan, S.S. (2011). Feeding aggregation of *Spratelloides delicatulus* on giant clams' gametes. *Coral Reefs*, 30, 167.

Madkour, H.A. (2005). Distribution and relationships of heavy metals in the giant clam (*Tridacna maxima*) and associated sediments from different sites in the Egyptian Red Sea Coast. *Egyptian Journal of Aquatic Research*, 31(2), 45–49.

Mohammed, T.A.A., Mohamed, E.M., Ebrahim, Y.M., Hafez, A.A. & Zamzamy, R.M.E. (2014). Some metal concentrations in the edible parts of *Tridacna maxima*, Red Sea, Egypt. *Environmental Earth Sciences*, 71(1), 301–309.

Morishima, S.-Y., Yamashita, H., O-hara, S., Nakamura, Y., Quek, V.Z., Yamauchi, M., et al. (2019). Study on expelled but viable zooxanthellae from giant clams, with an emphasis on their potential as subsequent symbiont sources. *PLOS ONE*, 14(7), e0220141.

Neo, M.L., Eckman, W., Vicentuan, K., Teo, S.L.-M. & Todd, P.A. (2015a). The ecological significance of giant clams in coral reef ecosystems. *Biological Conservation*, 181, 111–123.

Neo, M.L., Lee, B.Y., Vicentuan, K. & Todd, P.A. (2015b). Dichromatism in the commensal shrimp *Anchistus miersi* (De Man, 1888). *Marine Biodiversity*, 45(4), 877–878.

Roué, M., Darius, H., Chinain, M., Sibat, M. & Amzil, Z. (2016a). Ability of giant clams to bio-accumulate ciguatoxins from *Gambierdiscus* cells. *Harmful Algae News*, 55, 12–13.

Roué, M., Darius, H.T., Picot, S., Ung, A., Viallon, J., Gaertner-Mazouni, N., et al. (2016b). Evidence of the bioaccumulation of ciguatoxins in giant clams (*Tridacna maxima*) exposed to *Gambierdiscus* spp. cells. *Harmful Algae*, 57, 78–87.

Roué, M., Darius, H.T., Ung, A., Viallon, J., Sibat, M., Hess, P., et al. (2018). Tissue Distribution and Elimination of Ciguatoxins in *Tridacna maxima* (Tridacnidae, Bivalvia) Fed *Gambierdiscus polynesiensis*. *Toxins*, 10(5), 189.

Schmitt, W.L., McCain, J.C. & Davidson, E.S. (1973). *Crustaceorum Catalogus Pars 3. Decapoda I, Brachyura I, Fam. Pinnotheridae*. W. Junk: The Hague. 160pp.

Stauber, L.A. (1945). *Pinnotheres ostreum*, parasitic on the American Oyster, *Ostrea* (*gryphaea*) *virginica*. *The Biological Bulletin*, 88(3), 269–291.

Umeki, M., Yamashita, H., Suzuki, G., Sato, T., Ohara, S. & Koike, K. (2020). Fecal pellets of giant clams as a route for transporting Symbiodiniaceae to corals. *PLOS ONE*, 15(12), e0243087.

Vicentuan-Cabaitan, K., Neo, M.L., Eckman, W., Teo, S.L.-M. & Todd, P.A. (2014). Giant clam shells host a multitude of epibionts. *Bulletin of Marine Science*, 90(3), 795–796.

Ward, J.E., Rosa, M. & Shumway, S.E. (2019). Capture, ingestion, and egestion of microplastics by suspension-feeding bivalves: a 40-year history. *Anthropocene Coasts*, 2(1), 39–49.

Yeeting, B. (2009). Ciguatera-like fish poisoning from giant clams on Emao Island, Vanuatu. *SPC Fisheries Newsletter*, 129, 13–16.

Zhou, Z., Ni, X., Chen, S., Wu, Z., Tang, J., Su, Y., et al. (2022). Ingested microplastics impair the metabolic relationship between the giant clam *Tridacna crocea* and its symbionts. *Aquatic Toxicology*, 243, 106075.

5. CULTURAL AND SOCIOECONOMIC SIGNIFICANCE

Abd-Ebrah, N.A. & Peters, R.F. (2021). Giant Clam Conservation in Sabah: a need for the appreciation of the Bajau people's traditional ecological knowledge. *IOP Conference Series: Earth and Environmental Science*, 736(1), 012001.

Albert, D.D.A., Bujeng, V. & Chia, S. (2017). Traditional Shell Artefact Production in Northern Sabah. *Sabah Society Journal*, 33, 45–55.

Asato, S. (1991). The Distribution of *Tridacna* Shell Adzes in the Southern Ryukyu Islands. In: P. Bellwood (ed.). *Indo-Pacific Prehistory 1990, Volume 1. Bulletin of the Indo-Pacific Prehistory Association*, 10, 282–291.

Ayers, W.S. & Mauricio, R. (1987). Stone Adzes from Pohnpeian, Micronesia. *Archaeology in Oceania*, 22(1), 27–31.

Claus, C.A. (2017). Beyond Merroir: The Okinawan Taste for Clams. *Gastronomica*, 17(3), 49–57.

Heslinga, G.A. (1996). *Clams to Cash: How to Make and Sell Giant Clam Shell Products*. Center for Tropical and Subtropical Aquaculture Publication Number 125.

Hviding, E. (1993). *The rural context of giant clam mariculture in Solomon Islands: an anthropological study*. ICLARM Technical Report 39.

Kinoshita, N. (1992). A study of charm shells against the death evil: An archaeological and anthropological look at the shells of *Tridacna* family. *Comparative Folklore Studies*, 6, 5–39. [In Japanese]

Liu, C., Li, T., Liu, E., Li, C., Wang, A. & Gu, Z. (2019). Proximate Composition, Amino Acid Content, and Fatty Acid Profile of the Adductor Muscle and Mantle from Two Species of the Giant Clams *Tridacna crocea* and *Tridacna squamosa*. *Journal of Shellfish Research*, 38(3), 529–534.

Mahmoud, M.A.M., Zamzamy, R.M., Dar, M.A. & Mohammed, T.A.A. (2018). Biochemical assessment in the edible parts of *Tridacna maxima* Röding, 1798 collected from the Egyptian Red Sea. *The Egyptian Journal of Aquatic Research*, 44(3), 257–262.

Mies, M., Dor, P., Güth, A.Z. & Sumida, P.Y.G. (2017). Production in giant clam aquaculture: Trends and challenges. *Reviews in Fisheries Science & Aquaculture*, 25(4), 286–296.

Pavitt, A., Malsch, K., King, E., Chevalier, A., Kachelriess, D., Vannuccini, S., et al. (2021). *CITES and the sea: Trade in commercially exploited CITES-listed marine species*. FAO Fisheries and Aquaculture Technical Paper No. 666. Rome, FAO.

Reese, D.S. (1988). A New Engraved *Tridacna* Shell from Kish. *Journal of Near Eastern Studies*, 47(1), 35–41.

Reese, D.S. & Sease, C. (1993). Some Previously Unpublished Engraved *Tridacna* Shells. *Journal of Near Eastern Studies*, 52(2), 109–128.

Setiawan, H. (2013). Ancaman Terhadap Populasi Kima (*Tridacnidacna* sp.) dan Upaya Konservasinya di Taman Nasional Taka Bonerate. *Buletin Eboni*, 10(2), 137–147. [In Indonesian]

Tabugo, S.R.M., Pattuinan, J.O., Sespene, N.J.J. & Jamasali, A.J. (2013). Some economically important bivalves and gastropods found in the island of Hadji Panglima Tahil, in the province of Sulu, Philippines. *International Research Journal of Biological Sciences*, 2(7), 30–36.

Wabnitz, C., Taylor, M., Green, E. & Razak, T. (2003). *From Ocean to Aquarium: The global trade in marine ornamental species*. UNEP-WCMC, Cambridge, UK.

6. THREATS AND CHALLENGES

Addessi, L. (2001). Giant clam bleaching in the lagoon of Takapoto atoll (French Polynesia). *Coral Reefs*, 19, 220.

Andréfouët, S., Van Wynsberge, S., Gaertner-Mazouni, N., Menkes, C., Gilbert, A. & Remoissenet, G. (2013). Climate variability and massive mortalities challenge giant clam conservation and management efforts in French Polynesia atolls. *Biological Conservation*, 160, 190–199.

Andréfouët, S., Van Wynsberge, S., Kabbadj, L., Wabnitz, C.C.C., Menkes, C., Tamata, T., et al. (2018). Adaptive management for the sustainable exploitation of lagoon resources in remote islands: lessons from a massive El-Niño-induced giant clam bleaching event in Tuamotu atolls (French Polynesia). *Environmental Conservation*, 45(1), 30–40.

Apte, D., Narayana, S. & Dutta, S. (2019). Impact of sea surface temperature anomalies on giant clam population dynamics in Lakshadweep reefs: Inferences from a fourteen years study. *Ecological Indicators*, 107, 105604.

Armstrong, E.J., Duboussquet, V., Mills, S.C. & Stillman, J.H. (2020). Elevated temperature, but not acidification, reduces fertilization success in the small giant clam, *Tridacna maxima*. *Marine Biology*, 167, 8.

Brahmi, C., Chapron, L., Le Moullac, G., Soyez, C., Beliaeff, B., Lazareth, C.E., et al. (2019). Effects of temperature and pCO_2 on the respiration, biomineralization and photophysiology of the giant clam *Tridacna maxima*. *Conservation Physiology*, 9(1), coab041.

Bruno, J., Siddon, C., Witman, J., Colin, P. & Toscano, M. (2001). El Niño related coral bleaching in Palau, Western Caroline Islands. *Coral Reefs*, 20, 127-136.

Buck, B.H., Rosenthal, H. & Saint-Paul, U. (2002). Effect of increased irradiance and thermal stress on the symbiosis of *Symbiodinium microadriaticum* and *Tridacna gigas*. *Aquatic Living Resources*, 15, 107–117.

Chavanich, S., Viyakarn, V., Adams, P., Klammer, J. & Cook, N. (2012). Reef communities after the 2010 mass coral bleaching at Racha Yai Island in the Andaman Sea and Koh Tao in the Gulf of Thailand. *Phuket Marine Biological Center Research Bulletin*, 71, 103–110.

Enricuso, O.B., Conaco, C., Sayco, S.L.G., Neo, M.L. & Cabaitan, P.C. (2019) Elevated seawater temperatures affect the embryonic and larval development in the giant clam *Tridacna gigas* (Cardiidae: Tridacninae). *Journal of Molluscan Studies*, 85, 66–72.

Gomez, E. & Mingoa-Licuanan, S. (1998). Mortalities of giant clams associated with unusually high temperatures and coral bleaching. *Reef Encounter*, 24, 23.

Hester, F.J. & Jones, E.C. (1974). A survey of giant clams, Tridacnidae, on Helen Reef, a Western Pacific Atoll. *Marine Fisheries Review*, 36(7), 17–22.

Hirschberger, W. (1980). Tridacnid clam stocks on Helen Reef, Palau, Western Caroline Islands. *Marine Fisheries Review*, 42(2), 8–15.

Hviding, E. (1993). *The rural context of giant clam mariculture in Solomon Islands: an anthropological study*. ICLARM Technical Report 39.

Johnson, M.S., Prince, J., Brearley, A., Rosser, N.L. & Black, R. (2016). Is *Tridacna maxima* (Bivalvia: Tridacnidae) at Ningaloo Reef, Western Australia? *Molluscan Research*, 36(4), 264–270.

Junchompoo, C., Sinrapasan, N., Penpain, C. & Patsorn, P. (2013). Changing seawater temperature effects on giant clams bleaching, Mannai Island, Rayong Province, Thailand. In: *Proceedings of the Design Symposium on Conservation of Ecosystem (2013) (The 12th SEASTAR2000 workshop)* (pp. 71–76). Kyoto University Design School.

Larson, C. (2016). Shell trade pushes giant clams to the brink. *Science*, 351, 323–324.

Leggat, W., Buck, B.H., Grice, A. & Yellowlees, D. (2003). The impact of bleaching on the metabolic contribution of dinoflagellate symbionts to their giant clam host. *Plant, Cell and Environment*, 26, 1951–1961.

Li, J., Zhou, Y., Qin, Y., Wei, J., Shigong, P., Ma, H., et al. (2022). Assessment of juvenile vulnerability of symbiont-bearing giant clams to ocean acidification. *Science of the Total Environment*, 812, 152265.

Maboloc, M.A. & Villanueva, R.D. (2017). Effects of salinity variations on the rates of photosynthesis and respiration of the juvenile giant clam (Tridacna gigas, Bivalvia, Cardiidae). *Marine and Freshwater Behaviour and Physiology*, 50(4), 273–284.

Mekawy, M.S. & Madkour, H.A. (2012). Studies on the Indo-Pacific Tridacnidae (*Tridacna maxima*) from the Northern Red Sea, Egypt. *International Journal of GeoSciences*, 3, 1089–1095.

Mies, M. (2019). Evolution, diversity, distribution and the endangered future of the giant clam-Symbiodiniaceae association. *Coral Reefs*, 38, 1067–1084.

Militz, T.A., Kinch, J. & Southgate, P.C. (2015). Population demographics of *Tridacna noae* (Roding, 1798) in New Ireland, Papua New Guinea. *Journal of Shellfish Research*, 34(2), 329–335.

Munro, J.L. (1992). *Chapter 13: Giant clams*. FFA Report 92/75.

Neo, M.L., Erftemeijer, P.L.A., van Beek, J.K.L., van Maren, D.S., Teo, S.L.-M. & Todd, P.A. (2013). Recruitment constraints in Singapore's fluted giant clam (*Tridacna squamosa*) population - a dispersal approach. *PLOS ONE*, 8(3), e58819.

Pearson, R.G. (1977). Impact of foreign vessels poaching giant clams. *Australian Fisheries*, 36(7), 8-11, 23.

Ramah, S., Taleb-Hossenkhan, N., Todd, P.A., Neo, M.L. & Bhagooli, R. (2019). Drastic declines in giant clams (Bivalvia: Tridacninae) around Mauritius Island, Western Indian Ocean: implications for conservation and management. *Marine Biodiversity*, 49(2), 815–823.

Sayco, S.L.G., Conaco, C., Neo, M.L. & Cabaitan, P.C. (2019). Reduced salinities negatively impact fertilization success and early larval development of the giant clam *Tridacna gigas* (Cardiidae: Tridacninae). *Journal of Experimental Marine Biology and Ecology*, 516, 35–43.

Sayco, S.L.G., Cabaitan, P.C. & Kurihara, H. (2023). Bleaching reduces reproduction in the giant clam *Tridacna gigas*. *Marine Ecology Progress Series*, 706, 47–56.

Spencer, T., Teleki, K.A., Bradshaw, C. & Spalding, M.D. (2000). Coral bleaching in the Southern Seychelles during the 1997-1998 Indian Ocean Warm Event. *Marine Pollution Bulletin*, 40(7), 569–586.

Van Wynsberge, S., Andréfouët, S., Gaertner-Mazouni, N., Wabnitz, C.C.C., Gilbert, A., Remoissenet, G., et al. (2016). Drivers of density for exploited giant clam *Tridacna maxima*: a meta-analysis. *Fish and Fisheries*, 17(3), 567–584.

Watson, S.-A., Southgate, P.C., Miller, G.M., Moorhead, J.A. & Knauer, J. (2012). Ocean acidification and warming reduce juvenile survival of the fluted giant clam, *Tridacna squamosa*. *Molluscan Research*, 32(3), 177–180.

Watson, S.-A. (2015). Giant clams and rising CO_2: light may ameliorate effects of ocean acidification on a solar-powered animal. *PLOS ONE*, 10(6), e0128405.

Watson, S.-A. & Neo, M.L. (2021). Conserving threatened species during rapid environmental change: using biological responses to inform management strategies of giant clams. *Conservation Physiology*, 9, coab082.

White, M. (2019). *Initial assessment of a new coral bleaching event at Tongareva Atoll in the northern Cook Islands*. Hakono Hararanga Incorporated Report, January 2019.

Zhao, X. (2019). Life in the Wake of Hainan's Clam Shell Clampdown. Accessed in January 2023: https://www.sixthtone.com/news/1004054/life-in-the-wake-of-hainans-clam-shell-clampdown

7. CONSERVATION APPROACHES

Asaad, I., Lundquist, C.J., Erdmann, M.V. & Costello, M.J. (2017). Ecological criteria to identify areas for biodiversity conservation. *Biological Conservation*, 213(Part B), 309–316.

Basker, J.R. (1991). *Giant Clams in the Maldives – A stock assessment and study of their potential for culture*. BOBP/WP/72. Bay of Bengal Programme – Fisheries Resources.

Bell, J.D., Lane, I. & Hart, A.M. (1997). Culture, handling and air transport of giant clams from the South Pacific. In: B. Paust & J.B. Peters (eds.). *Marketing and Shipping Live Aquatic Products '96* (pp. 60–66). Northeast Regional Agricultural Engineering Service, New York.

Bell, J.D. (1999). Reducing the costs of restocking giant clams in Solomon Islands. *Coral Reefs*, 18, 326.

Bell, J.D., Munro, J.L., Nash, W.J., Rothlisberg, P.C., Loneragan, N.R., Ward, R.D., et al. (2005). Chapter 2. Restocking Initiatives - 2.1. Giant Clams. In: A.J. Southward, C.M. Young & L.A. Fuiman (eds.). *Restocking and Stock Enhancement of Marine Invertebrate Fisheries* (pp. 10–41). Advances in Marine Biology, Volume 49.

Benzie, J.A.H. & Williams, S.T. (1996). Limitations in the genetic variation of hatchery produced batches of the giant clam, *Tridacna gigas*. *Aquaculture*, 139, 225–241.

Braley, R.D. & Muir, F. (1995). The Case History of a Large Natural Cohort of the Giant Clam *Tridacna gigas* (Fam. Tridacnidae) and the Implications for Re-stocking Depauperate Reefs with Maricultured Giant Clams. *Asian Fisheries Science*, 8, 229–237.

Braley, R.D. (1992). *The Giant Clam: Hatchery and Nursery Culture Manual*. ACIAR Monograph No. 15.

Braley, R.D. (1996). The importance of aquaculture and establishment of reserves for the restocking of giant clams on over-harvested reefs in the Indo-Pacific region. In: T.G. Heggberget (ed.). *The Role of Aquaculture in World Fisheries, Proceedings of the World Fisheries Congress, Theme 6* (pp. 136-147). New Delhi: Oxford and IBH Publishing.

Braley, R.D., Militz, T.A. & Southgate, P.C. (2018). Comparison of three hatchery culture methods for the giant clam *Tridacna noae*. *Aquaculture*, 495, 881–887.

Cabaitan, P.C. & Conaco, C. (2017). Bringing back the giants: juvenile *Tridacna gigas* from natural spawning of restocked giant clams. *Coral Reefs*, 36(2), 519.

Cadotte, M.W. & Tucker, C.M. (2018). Difficult decisions: Strategies for conservation prioritization when taxonomic, phylogenetic and functional diversity are not spatially congruent. *Biological Conservation*, 225: 128–133.

Clifton, J. (2009). Science, funding and participation: key issues for marine protected area networks and the Coral Triangle Initiative. *Environmental Conservation*, 36(2), 91–96.

Coastal Fisheries Species Regulations, Management Plans and Legislation in Pacific Island Countries and Territories, SPC Coastal Fisheries Programme (Online). Accessed in January 2023: http://www.spc.int/CoastalFisheries/CFM/LegalText/ByJurisdiction

Collar, N.J. (1996). The reasons for Red Data Books. *Oryx*, 30(2), 121–130.

Copland, J.W. & Lucas, J.S. (eds.) (1988). *Giant clams in Asia and the Pacific.* ACIAR Monograph No. 9.

Crawford, C.M., Lucas, J.S. & Munro, J.L. (1987). The Mariculture of Giant Clams. *Interdisciplinary Science Reviews*, 12(4), 333–340.

Cullen, R. (2012). Biodiversity protection prioritisation: a 25-year review. *Wildlife Research*, 40(2), 108–116.

Davila, F., Sloan, T., Milne, M. & van Kerkhoff, L. (2017). *Impact assessment of giant clam research in the Indo-Pacific region.* ACIAR Impact Assessment Series Report No. 94.

Ellis, S. (1998). *Spawning and early larval rearing of giant clams (Bivalvia: Tridacnidae).* Center for Tropical and Subtropical Aquaculture Publication No. 130.

Ezekiel, A. (2018). *Review of Marine Wildlife Protection Legislation in ASEAN TRAFFIC.* Petaling Jaya, Selangor, Malaysia. Accessed in January 2023: https://www.traffic.org/site/assets/files/11344/marine-wildlife-protection-legislation-in-asean.pdf

Feltham, J. & Capdepon, L. (2021). *Giant Clam Shells, Ivory, and Organised Crime: Analysis of a potential new nexus.* Wildlife Justice Commission. Accessed in January 2023: https://wildlifejustice.org/wp-content/uploads/2021/10/Giant-Clam-Shells-Ivory-And-Organised-Crime_A-Potential-New-Nexus_WJC_spreads.pdf

Fitt, W.K., Rees, T.A.V., Braley, R.D., Lucas, J.S. & Yellowlees, D. (1993). Nitrogen flux in giant clams: size dependency and relationship to zooxanthellae density and clam biomass in the uptake of dissolved inorganic nitrogen. *Marine Biology*, 117, 381–386.

Gomez, E.D. & Mingoa-Licuanan, S.S. (2006). Achievements and lessons learned in restocking giant clams in the Philippines. *Fisheries Research*, 80, 46–52.

Gomez, E.D. (2015). Rehabilitation of biological resources: coral reefs and giant clam populations need to be enhanced for a sustainable marginal sea in the Western Pacific. *Journal of International Wildlife Law and Policy*, 12(2), 120–127.

Heslinga, G.A., Perron, F.E. & Orak, O. (1984). Mass culture of giant clams (F. Tridacnidae) in Palau. *Aquaculture*, 39(1-4), 197–215

Heslinga, G.A. (1991). *History and current status of the MMDC giant clam project*. A Special Report Prepared for: The House of Delegates, Third Olbiil Era Kelulau (Palau National Congress). February 10, 1991. Koror, Palau: Micronesian Mariculture Demonstration Center.

Heslinga, G.A. (2012). The Origin and Future of Farming Giant Clams. *The Reef and Marine Aquarium Magazine*, 9(6), 38–52.

Heslinga, G.A. (2013). *Saving Giants (eBook): Cultivation and Conservation of Tridacnid Clams*. Kailua-Kona, Hawaii: Indo-Pacific Sea Farms. https://www.blurb.com/b?ebook=374835

Iwai, K., Kiso, K. & Kubo, H. (2006). Biology and Status of Aquaculture for Giant Clams (Aquaculture) in the Ryukyu Islands, Southern Japan. In: J.H. Primavera, E.T. Quinitio & M.R.R. Eguia (eds.). *Proceedings of the Regional Technical Consultation on Stock Enhancement for Threatened Species of International Concern, Iloilo City, Philippines, 13-15 July 2005* (pp. 27–38). Aquaculture Department, Southeast Asian Fisheries Development Center.

IUCN (2012). *IUCN Red List Categories and Criteria: Version 3.1. Second edition*. Gland, Switzerland and Cambridge, UK: IUCN.

IUCN (2015). In: L.M. Bland, D.A. Keith, N.J. Murray & J.P. Rodriguez (eds.). *Guidelines for the application of IUCN Red List of Ecosystems Categories and Criteria, Version 1.0*. IUCN.

IUCN (2022). The IUCN Red List of Threatened Species. Version 2022-1. Accessed in January 2023: https://www.iucnredlist.org

Kakuma, S. & Kamimura, M. (2011). Okinawa: Effective conservation practices from Satoumi in a coral reef ecosystem. In: United Nations University Institute of Advanced Studies Operating Unit Ishikawa/Kanazawa (ed.). *Biological and cultural diversity in coastal communities: Exploring the potential of Satoumi for implementing the ecosystem approach in the Japanese Archipelago* (pp. 86–93). Montreal, QC, Canada: Secretariat of the Convention on Biological Diversity. CBD Technical Series No. 61.

Kakuma, S. (2022). Satoumi systems: promoting integrated coastal resources managements: an empirical review. *Sustainability*, 14, 11702.

Kinch, J. (2001). Clam Harvesting, the Convention on the International Trade in Endangered Species (CITES) and Conservation in Milne Bay Province, Papua New Guinea. *SPC Fisheries Newsletter*, 99, 24–36.

Kinch, J. & Teitelbaum, A. (2010). *Proceedings of the Regional Workshop on the Management of Sustainable Fisheries for Giant Clams (Tridacnidae) and CITES capacity building*. SPC Aquaculture Technical Papers.

Lucas, J.S. (1994). The biology, exploitation, and mariculture of giant clams (Tridacnidae). *Reviews in Fisheries Science*, 2(3), 181–223.

Lucas, J.S. (1997). Giant clams: mariculture for sustainable mariculture. In: M. Bolton (ed.). *Conservation and the Use of Wildlife Resources* (pp. 77–95). Chapman & Hall.

Lyons, Y., Cheong, D., Neo, M.L. & Wong, H.F. (2018). Managing giant clams in the South China Sea. *The International Journal of Marine and Coastal Law*, 33, 467–494.

Moore, D. (2022) Farming Giant Clams in 2021: A Great Future for the 'Blue Economy' of Tropical Islands. In: D. Moore et al. (eds.). *Aquaculture: Ocean Blue Carbon Meets UN-SDGS* (pp. 131–153). Sustainable Development Goals Series.

Moorhead, A. (2018). Giant clam aquaculture in the Pacific region: perceptions of value and impact. *Development in Practice*, 28(5), 624–635.

Mies, M., Dor, P., Güth, A.Z. & Sumida, P.Y.G. (2017). Production in giant clam aquaculture: Trends and challenges. *Reviews in Fisheries Science & Aquaculture*, 25(4), 286–296.

Munro, J.L. & Heslinga, G.A. (1983). Prospects for the commercial cultivation of giant clams (Bivalvia: Tridacnidae). *Proceedings of the Annual Gulf Caribbean Fisheries Institute*, 35, 122–134.

Murakoshi, M. (1986). Farming of the boring giant clam, *Tridacna crocea* Lamarck. *Galaxea*, 5, 239–254.

Neo, M.L. & Todd, P.A. (2012). Giant clams (Mollusca: Bivalvia: Tridacninae) in Singapore: history, research and conservation. *Raffles Bulletin of Zoology*, 25, 67–78.

Neo, M.L., Wabnitz, C.C.C., Braley, R.D., Heslinga, G.A., Fauvelot, C., Van Wynsberge, S., et al. (2017). Giant Clams (Bivalvia: Cardiidae: Tridacninae): A Comprehensive Update of Species and Their Distribution, Current Threats and Conservation Status. *Oceanography and Marine Biology: An Annual Review*, 55, 87–388.

Neo, M.L. (2020). Conservation of Giant Clams (Bivalvia: Cardiidae). In: M.I. Goldstein & D.A. DellaSala (eds.). *Encyclopedia of the World's Biomes, Volume 4* (pp. 527–538). Elsevier.

Ng, L.W.K., Chisholm, C., Carrasco, L.R., Darling, E.S., Guilhaumon, F., Mooers, A.Ø., et al. (2022). Prioritizing phylogenetic diversity to protect functional diversity of reef corals. *Diversity and Distributions*, 28(8), 1721–1734.

Pavitt, A., Malsch, K., King, E., Chevalier, A., Kachelriess, D., Vannuccini, S., et al. (2021). *CITES and the sea: Trade in commercially exploited CITES-listed marine species*. FAO Fisheries and Aquaculture Technical Paper No. 666. Rome, FAO.

Pimiento, C., Leprieur, F., Silvestro, D., Lefcheck, J.S., Albouy, C., Rasher, D.B., et al. (2020). Functional diversity of marine megafauna in the Anthropocene. *Science Advances*, 6(16), eaay7650.

Redding, D.W. & Mooers, A.Ø (2006). Incorporating Evolutionary Measures into Conservation Prioritization. *Conservation Biology*, 20(6), 1670–1678.

Rodrigues, A.S.L., Pilgrim, J.D., Lamoreux, J.F., Hoffmann, M. & Brooks, T.M. (2006). The value of the IUCN Red List for conservation. *Trends in Ecology & Evolution*, 21(2), 71–76.

Sulaiman, Z.H. (2000). Marine Life of Brunei Darussalam. *Bruneiana: Anthology of Science Articles*, 88–91.

Teitelbaum, A. & Friedman, K. (2008). Successes and failures in reintroducing giant clams in the Indo-Pacific region. *SPC Trochus Information Bulletin*, 14, 19–26.

Tisdell, C. (ed.) (1992). *Giant Clams in the Sustainable Development of the South Pacific: socioeconomic issues in mariculture and conservation*. ACIAR Monograph No. 18.

Tucker, C.M., Davies, T. J., Cadotte, M.W. & Pearse, W.D. (2018). On the relationship between phylogenetic diversity and trait diversity. *Ecology*, 99(6), 1473–1479.

Tucker, C.M., Aze, T., Cadotte, M.W., Cantalapiedra, J.L., Chisholm, C., Díaz, S., et al. (2019). Assessing the utility of conserving evolutionary history. *Biological Reviews*, 94(5), 1740–1760.

UNEP-WCMC (2012). *Review of species subject to long-standing positive opinions: species other than corals and butterflies from Asia and Oceania*. UNEP-WCMC, Cambridge.

Vellend, M., Cornwell, W.K., Magnuson-Ford, K., & Mooers, A.Ø. (2011). Measuring phylogenetic biodiversity. In: A.E. Magurran & B.J. McGill (eds.), *Biological diversity: Frontiers in measurement and assessment* (pp. 194–207). Oxford University Press.

Villéger, S., Mason, N.W.H. & Mouillot, D. (2008). New multidimensional functional diversity indices for a multifaceted framework in functional ecology. *Ecology*, 89(8), 2290–2301.

Waters, C.G., Story, R. & Costello, M.J. (2013). A methodology for recruiting a giant clam, *Tridacna maxima*, directly to natural substrata: A first step in reversing functional extinctions? *Biological Conservation*, 160, 19–24.

Wells, S. (1997). *Giant clams: Status, trade and mariculture, and the roles of CITES management*. IUCN, Gland, Switzerland and Cambridge, UK. 77pp.

Wells, S.M., Pyle, R.M. & Collins, N.M. (1983). *The IUCN Invertebrate Red Data Book*. IUCN, Gland, Switzerland. 632pp.

9. GENUS *HIPPOPUS*, 10. GENUS *TRIDACNA*, 11. DICHOTOMOUS KEY

Bonfitto, A., Sabelli, B., Tommasini, S. & Herbert, D. (1994). Marine molluscan taxa from Mozambique described by G.G. Bianconi and preserved in the Zoological Museum of the University of Bologna. *Annals of the Natal Museum*, 35(1), 133–138.

Borsa, P., Fauvelot, C., Andrefouet, S., Chai, T.-T., Kubo, H. & Liu, L.-L. (2015). On the validity of Noah's giant clam *Tridacna noae* (Röding, 1798) and its synonymy with Ningaloo giant clam *Tridacna ningaloo* Penny & Willan, 2014. *Raffles Bulletin of Zoology*, 63, 484–489.

Fauvelot, C., Zuccon, D., Borsa, P., Grulois, D., Magalon, H., Riquet, F., et al. (2020). Phylogeographical patterns and a cryptic species provide new insights into Western Indian Ocean giant clams phylogenetic relationships and colonization history. *Journal of Biogeography*, 47(5), 1086–1105.

Huber, M. & Eschner, A. (2010). *Tridacna (Chametrachea) costata* Roa-Quiaoit, Kochzius, Jantzen, Al-Zibdah & Richter from the Red Sea, a junior synonym of *Tridacna squamosina* Sturany, 1899 (Bivalvia, Tridacnidae). *Annalen des Naturhistorischen Museums in Wien. Serie B für Botanik und Zoologie*, 112, 153–162.

Lucas, J.S., Ledua, E. & Braley, R.D. (1991). *Tridacna tevoroa* Lucas, Ledua and Braley: a recently-described species of giant clam (Bivalvia; Tridacnidae) from Fiji and Tonga. *The Nautilus*, 105(3), 92–103.

Monsecour, K. (2016). A New Species of Giant Clam (Bivalvia: Cardiidae) from the Western Indian Ocean. *Conchylia*, 46, 69–77.

Richter, C., Roa-Quiaoit, H., Jantzen, C., Al-Zibdah, M. & Kochzius, M. (2008). Collapse of a New Living Species of Giant Clam in the Red Sea. *Current Biology*, 18(17), 1349–1354.

Rosewater, J. (1965). The family Tridacnidae in the Indo-Pacific. *Indo-Pacific Mollusca*, 1, 7–82.

Rosewater, J. (1982). A new species of *Hippopus* (Bivalvia: Tridacnidae). *The Nautilus*, 96, 3–6.

Sirenko, B.I. & Scarlato, O.A. (1991). *Tridacna rosewateri* sp. n. A new species of giant clam from the Indian Ocean. *La Conchiglia*, 22, 4–9.

Su, Y., Hung, J.-H., Kubo, H. & Liu, L.-L. (2014). *Tridacna noae* (Röding, 1798) - a valid giant clam species separated from *T. maxima* (Röding, 1798) by morphological and genetic data. *Raffles Bulletin of Zoology*, 62, 124–135.

Sundy, R., Kaullysing, D. & Bhagooli, R. (2021). First *in-situ* observation of the endemic giant clam Tridacna rosewateri from the Nazareth Bank, Mascarene Plateau. *Western Indian Ocean Journal of Marine Science*, (No. 2/2021 (2021): Special Issue-Studies on the Mascarene Plateau), 193–196.

12. GIANT CLAM IN VERNACULAR LANGUAGES

Anam, R. & Mostarda, E. (2012). *Field identification guide to the living marine resources of Kenya*. FAO Species Identification Guide for Fishery Purposes. Rome, FAO.

Bao, K. & Drew, J. (2016). Traditional ecological knowledge, shifting baselines, and conservation of Fijian molluscs. *Pacific Conservation Biology*, 23(1), 81–87.

Dalzell, P., Lindsay, S.R. & Patiale, H. (1993). *Fisheries resources survey of the island of Niue*. South Pacific Commission.

Elbert, S.H. (1975). *Dictionary of the language of Rennell and Bellona. Part 1: Rennellese and Bellonese to English*. National Museum of Denmark: Copenhagen.

Govan, H. (1989). Vernacular names of tridacnid clams in the Pacific region. *Aquabyte - ICLARM Newsletter of the Network of Tropical Aquaculture Scientists*, 2(1): 5–7.

Job, S. & Ceccarelli, D. (2012). *Tuvalu Marine Life - An Alofa Tuvalu Project with the Tuvalu Fisheries Department and Funafuti, Nanumea, Nukulaelae Kaupules*. Scientific Report, December 2012.

Kanno, K., Kotaki, Y. & Yasumoto, T. (1976). Distribution of toxins in molluscs associated with coral reefs. *Bulletin of the Japanese Society of Scientific Fisheries*, 42(12), 1395–1398.

Neo, M.L. & Todd, P.A. (2012). Giant clams (Mollusca: Bivalvia: Tridacninae) in Singapore: History, Research and Conservation. *The Raffles Bulletin of Zoology*, 25: 67–78.

Psomadakis, P.N., Htun Thein, Russell, B.C. & Mya Than Tun (2019). *Field identification guide to the living marine resources of Myanmar. FAO Species Identification Guide for Fishery Purposes*. Rome, FAO and MOALI.

Thomas, F.R. (2001). Mollusk Habitats and Fisheries in Kiribati: An Assessment from the Gilbert Islands. *Pacific Science*, 55(1), 77–97.

Wakum, A., Takdir, M. & Talakua, S. (2017). Jenis-Jenis Kima dan Kelimpahannya di Perairan Amdui Distrik Batanta Selatan Kabupaten Raja Ampat. *Jurnal Sumberdaya Akuatik Indopasifik*, 1(1), 43–52. [In Indonesian]

Zann, L. (1989). *A preliminary check list of the major species of fishes and other marine organisms in Western Samoa (Samoan/Scientific/English)*. FAO/UNDP SAM/89/002.

www.ingramcontent.com/pod-product-compliance
Lightning Source LLC
Chambersburg PA
CBHW050601190326
41458CB00007B/2137